Altered States of Consciousness

Altered States of Consciousness

Experiences Out of Time and Self

Marc Wittmann

translated by Philippa Hurd

The MIT Press
Cambridge, Massachusetts
London, England

First MIT Press paperback edition, 2023
This translation © 2018 Massachusetts Institute of Technology

Originally published as *Wenn die Zeit stehen bleibt* by Marc Wittmann, © Verlag C.H.Beck oHG, München 2015

This book was set in ITC Stone Serif Std by Toppan Best-set Premedia Limited. Printed and bound in the United States of America.

Library of Congress Cataloging-in-Publication Data

Names: Wittmann, Marc, author.
Title: Altered states of consciousness : experiences out of time and self / Marc Wittmann ; translated by Philippa Hurd.
Other titles: Wenn die Zeit stehen bleibt. English
Description: Cambridge, MA : MIT Press, [2018] | Includes bibliographical references and index.
Identifiers: LCCN 2017055646 | ISBN 9780262038317 (hardcover : alk. paper)—9780262546089 (paperback)
Subjects: LCSH: Time perception. | Time--Psychological aspects. | Altered states of consciousness.
Classification: LCC BF468 .W57313 2018 | DDC 153.7/53--dc23 LC record available at https://lccn.loc.gov/2017055646

10 9 8 7 6 5 4 3

For Oksana

Contents

Prologue: An "I" Awakes

I remember waking up one morning. For a moment I didn't know who I was or where I was. I was. That much I remember. But I didn't know who was waking up. This feeling of conscious experience persisted for just a felt moment. Then I awoke fully, accessed my memories, and knew once again who I am and where I was.

The experience of waking up in darkness and not knowing where you are is not that uncommon. Still drowsy, we find ourselves in a past period of our lives, perhaps in the bed and the bedroom of our youth. One time I woke up in my bedroom in Freiburg and for a few seconds I was convinced I was in San Diego, where I had lived a few years before. I knew about the position of the bed in the San Diego house, I felt the presence of the spatial position of the window and the street outside and, as I awoke, I expected to find myself there, so real was the impression—for a short time I was my former San Diego self. But I awoke in my apartment in Freiburg.

Occasionally it happens that it takes a few seconds to figure out your life as a whole and the space you're in. So far, however, it has only happened to me one time that I didn't know *who* I was. I searched for possible memories, but they wouldn't update. For a moment I was divested of my self. Until I became aware of myself again—in the form of knowledge about myself, as the memory of who I am. The experience lasted for only a brief moment, and it wasn't frightening. By contrast, the Swedish poet Tomas Tranströmer experienced his loss of self on waking as extremely stressful:

The Name

I grow sleepy during the car journey and I drive in under the trees at the side of the road. I curl up in the back seat and sleep. For how long? Hours. Dusk has fallen.

Suddenly I'm awake and don't know where I am. Wide awake, but it doesn't help. Where am I? WHO am I? I am something that wakens in a back seat, twists about in panic like a cat in a sack. Who?

At last my life returns. My name appears like an angel. Outside the walls a trumpet signal blows (as in the *Leonora* Overture) and the rescuing footsteps come down the overlong stairway. It is I! It is I!

But impossible to forget the fifteen-second struggle in the hell of oblivion, a few meters from the main road, where the traffic drives past with its lights on.[1]

Where am I? Who am I? The above experience shatters the foundations of our everyday understanding of our self. I am in the world and experience the world. Of course I know that I may be deluded about the circumstances of the world. Perceptual illusions exist. As René Descartes showed,[2] I can be hallucinating or dreaming, I can be led to believe in a world, but it is surely I who is hallucinating or dreaming. It is I who is deluding myself or being deluded. In this I am always certain of my self.

The example of waking up demonstrates that the memory, the autobiographical remembrance, must exist in order that I know who I am. For example, for Marcel Proust in his *In Search of Lost Time*, memory is constitutive of the self and the felt life as a whole. In the novel, the search for lost time is ultimately fulfilled through the comprehensive and detailed remembering of events in the life of the first-person narrator, who is writing his life story. Time is regained through the remembered and recorded life. For Proust too, one source of this idea is the transition between sleeping and waking. Proust's narrator recounts the experience of a short-term loss of the self that is resolved only by an influx of memories:

> When I awoke in the middle of the night, not knowing where I was, I could not even be sure at first who I was; I had only the most rudimentary sense of existence, such as may lurk and flicker in the depths of an animal's consciousness; I was more destitute than the cave-dweller; but then the memory—not yet of the place in which I was, but of various other places where I had lived and might now very possibly be—would come down like a rope let down from heaven to draw me up out of the abyss of not-being, from which I could never have escaped by myself: in a flash I would traverse centuries of civilisation, and out of a blurred glimpse of oil-lamps, then of shirts with turned-down collars, would gradually piece together the original components of my ego.[3]

According to this account, the self forms out of memory. This is often described as the narrative self that creates itself from the stories we tell about ourselves. But there is also a sense of self that exists as "the mere feeling of being" without autobiographical knowledge. This is minimal self or "core self."[4] In the seconds of waking up, as the narrative self is not updating, consciousness is focused on something nevertheless: it is the physical self that is at the center of perception and thought, which enables the differentiation between the self and non-self. Under normal circumstances we are aware of our experiences, memories, and expectations, the objects of our consciousness. Below the surface, however, we also have a minimal self, the egocentric anchor of all experiences that in the above-mentioned situation of memoryless awakening is suddenly experienced very clearly, as the usual objects of our consciousness, perceptions, and memories are missing. I am thrown back upon myself.[5]

In such a case the experience of self can be understood as an "ego-pole." My "ego-subject" is focused on an "ego-object": I perceive myself. However, there is a fundamental problem here, as the ego-object is categorically different from the ego-subject. If we observe ourselves self-referentially—that is, the ego-subject observes itself—it always observes itself as an ego-object. In the philosophy of subjectivity in the nineteenth century, this problem brought about major bodies of thought. The impossibility of being able to perceive oneself as an ego-subject was described effectively by Thomas Bernhard in his novella *Walking*:

> If we observe ourselves, we are never observing ourselves but someone else. Thus we can never talk about self-observation, or when we talk about the fact that we observe ourselves we are talking as someone we never are when we are not observing ourselves, and thus when we observe ourselves we are never observing the person we intended to observe but someone else. The concept of self-observation and so, also, of self-description is thus false.[6]

These experiences on waking up, the awakening of the self, can be quite extraordinary as states of consciousness. Marcel Proust describes major alterations in temporal perception: "In a flash I would traverse centuries of civilisation." Altered states of consciousness very often go hand in hand with an altered perception of space and time, as we shall see. Ultimately our perception and our thoughts are organized in terms of space and time. Extraordinary states of consciousness must therefore also affect space and time. When I wake up, I look drowsily at the clock. I go back to sleep and have a very striking and complex dream full of drama, which, I feel, goes on for a long time. When I wake up again, I look at the clock to see that only three minutes have passed. Subjectively, in retrospect I remember impressions that are the length of a movie. Subjective and objective time differ dramatically from one another.

In the transition from sleeping to waking we experience the boundaries of our usual state of self.[7] Every time we wake we become conscious of our selves once again; we are inserted into our state of awakeness. But in isolated cases the process of becoming conscious does not happen seamlessly—the ego does not recognize itself. Through such moments we have the opportunity to investigate the enigma of consciousness, revealing how the conscious self depends on factors yet to be determined, which are constitutive of self-consciousness.

Through these and other phenomena, in extraordinary states of consciousness that deviate strikingly from normal experience we gain access to one of the major questions: what is consciousness and self-consciousness. Here, the question of how subjective time arises is of particular importance. Subjective time and consciousness, felt time, and experience of self are closely related:[8] I am my time; through my experience of self I reach a feeling of time. If we have a better understanding of the subjective experience of time, then important aspects of self-consciousness will also have been understood better. Scientific research into extraordinary states of consciousness provides an avenue toward understanding what constitutes consciousness, by way of exploring alterations in the experience of time.

The subject of this book is the close relationship between consciousness of self and consciousness of time. In extraordinary states of consciousness—moments of shock, meditation, sudden mystical experiences, near-death experiences, under the influence of drugs—temporal consciousness is fundamentally altered. Hand in hand with this goes an altered consciousness of space and self. In these extreme circumstances, time and concepts of space and self are modulated together—intensified or weakened together. But in more ordinary situations, too, such as boredom, the experience of flow, and idleness, time and self are collectively altered. Ultimately, clinical research in psychiatry and neurology, as well as studies in basic brain research, show how the concepts of body, self, space, and time are linked. All these phenomena and conditions will be discussed here. The empirical findings and conceptual insights—which have never before been brought together in this way—are astonishing but will not surprise those familiar with philosophical ideas. Some philosophical systems of ideas, to which we will refer briefly in the relevant places, have unified the aforementioned components. The latest research results in psychology and the neurosciences can be viewed in relation to these philosophical ideas. Through the conceptual understanding of the meaning of time and how it is modulated in extreme moments of life we can find out more about what consciousness is—about who we are.

1 On Time Consciousness

Time Expansion and Moments of Terror

As a matter of course we perceive our world through its spatial and temporal characteristics. The world expands spatially in its three dimensions; equally, our experiences take place in a temporal system of future (expected), present (lived), and past (remembered) events. As our everyday observations show, the felt passing of time is not always the same. Ten boredom-filled minutes can go by painfully slowly, while ten minutes full of joy and entertainment fly by in an instant. If we look at the situations when time expansion and acceleration occur, we can see that the extent to which our attention is focused on time determines how the passage of time is perceived. During a period of waiting without any distractions, during which nothing much of interest happens, we are particularly aware of the time and it passes slowly. We might think of some tough lessons at school when we were young, or waiting for the phone call from a loved one that doesn't come. Our attention is completely focused on the time—which won't go by. By contrast, if we are completely absorbed in an activity, in reading, playing, or exciting work, we don't pay attention to time and it passes quickly. In the experience of flow—in particular, being completely immersed in a challenging activity, which we enjoy but which also demands all our skills—we have the feeling that time is moving quickly on.[1] In such experiences we are removed from time and our selves and don't notice time passing; in extreme cases, hours go by like minutes.

Alongside our attention, our physical state also has an effect on the present experience of time. In situations of increased emotional agitation, it often feels as if time is expanding. If we are very angry or in the terrifying moments of an accident, but also when we experience the greatest happiness, time seems to slow down: the last steps to the top of a mountain after considerable effort and as the panoramic view appears; the moment before the longed-for kiss when it becomes clear that it's about to happen. These are experiences of the greatest emotional presence: our attention is heightened and the body enters a state of extreme preparedness. For a moment, time almost stands still.

This leads us to ask the question of what we're paying attention to when we say we're paying attention to the time. Ultimately there's no sense organ for the perception of time, as the ears perceive sounds or the eyes sight. Nevertheless we often perceive time quite directly, sometimes even as unpleasant, for example when we are bored and time simply won't go by. It makes complete sense to talk about the perception of time, although there is no dedicated sense organ to do this. How does the experience of the passage of time arise? Our temporal feeling for the current passing of time, as we experience it elapsing quickly or slowly, is based on our bodily experience.[2] Moments of strong, emotional excitement expand subjective time, as bodily states are experienced either directly or as mediated by the emotions. But even in situations of relative calm, time can expand. During a period of waiting, I am left to my own devices and I experience myself and my physicality more intensively. Practicing forms of meditation, which consciously draw attention to the body, gives rise to the feeling of time passing slowly. Novices in meditation, in particular, feel time expansion very clearly if they have to spend 30 minutes without moving. Physicality becomes tangible and time passes "infinitely" slowly. If I pay attention to the time of the present moment, I feel the passage of time particularly intensively across my body, which is me. I am time.

Reports of extraordinary states of consciousness, in particular, bear convincing witness to the mutability of the feeling of time.[3] Having lived through dangerous situations, people often report that the events took place in slow motion, that a moment that was actually brief expanded noticeably. These days, in our motorized society, this kind of experience often pertains to a traffic accident or to a narrow escape in a dangerous road situation. John C. Eccles, the neurophysiologist and Nobel laureate, vividly described such an altered experience of time; his account is remarkable for the fact that his wife, who sat alongside him in the car, had the same experience:

> Down that dark road there was a dark red truck hurtling at perhaps 80 mph down the hill. My wife and I didn't see it until it flashed out of the darkness into the light. It was too late to stop so that all we could do was to try to accelerate to get out, and we were moving slowly because we had just come off a standing start! As I watched this truck coming closer and closer, time seemed to go on forever. [...] Then in the end miraculously, I found out that the back of the car even wasn't hit and the truck moved past, but all in slow motion. It was the most incredible experience, and my wife had the same experience that time had almost come to a stop in this emergency.[4]

Tomas Tranströmer, the poet and Nobel laureate in literature, underwent a similar experience. In his poem *Alone* (excerpted below[5]), he tells of experiencing a moment of terror in which, from one moment to the next, he is faced with a possible accident:

...

One evening in February I almost died here.
The car skidded sideways on the ice, but
onto the wrong side of the road. The approaching cars—
their lights—closed in.

...

The approaching traffic had huge lights.
They shone on me while I pulled at the wheel
in a transparent terror that floated like egg white.
The seconds grew—there was space in them—
they grew as big as hospital buildings.
You could almost pause
and breathe for a while
before being crushed.

...

Such experiences of events proceeding in slow motion are, of course, not the preserve of Nobel laureates; they can happen to all those who are in states of extreme stress where their lives are directly threatened. In a systematic study of a large number of accounts given by survivors after life-threatening situations such as car accidents, but also in cases of near-drowning or falling, the phenomenon of the "altered passage of time" applied in 71 percent of all cases.[6] In most cases, a subjective deceleration of time and overestimation of the duration of the event was reported. An early systematic study of near-death experiences was published in 1892 by Albert Heim, a Swiss geologist, who had himself undergone an important experience, when he survived a fall on Säntis, a mountain in the Appenzell region of Switzerland.[7] During the fall, he had a great many thoughts and experiences that didn't seem to be happening in such a short span of time. His entire past life flashed by him in pictures; he felt surrounded by a heavenly light. Here, Albert Heim was having a typical near-death experience (actually a fear-of-death experience), as described in accounts given by people of all ages and cultures. We'll return to this subject in more detail later on.

In a 2012 article, the Finnish philosopher Valtteri Arstila analyzed accounts of life-threatening accidents and highlighted at least five components that are essential to the phenomenon of altered time consciousness:[8]

1. The person feels that the duration of events is expanding and the temporal progression of events is slowed down.

2. There is an enhanced mental sharpness and increased speed of thoughts.

3. Those affected act quickly and purposefully.

4. Their attention is focused on aspects of the situation that are essential for survival.

5. Visual and auditory impressions are unusually sharp.

According to Arstila's observations, in moments of the greatest danger a survival reaction is triggered in the entire body—the fight-or-flight response, an extremely heightened organismic stimulation: attention is sharpened, sense impressions display a never-before-experienced intensity, and thinking and acting speed up. In the brain (as Arstila summarizes the findings of brain research in his article), these physical survival reactions go hand in hand with, among other things, the release of the neurotransmitter noradrenalin, which is produced in the *locus coeruleus* in the pons of the brain stem. The noradrenergic neural pathways emanating from the locus coeruleus innervate specific areas of the brain that are involved in focused attention and working memory, as well as speed and accuracy when taking emergency action. These processes are crucial with regard to the altered perception of time, to the extent that a faster processing of information in the observer's organism leads to the external processes of the environment being experienced as relatively decelerated, resulting in the slow-motion effect.

After experiencing dangerous situations, many people report an acceleration of thought processes, the sign of an increased processing speed in the entire body while, in relation to this, external events seem to slow down. This heightened level of arousal might enable faster reactions in order to escape the dangerous situation more efficiently—for example hitting the gas pedal faster or turning the steering wheel at the right moment. In fact, people report that they acted faster and more expediently than normal in these dangerous situations, in order to escape danger. This means that the relative balance between internal, physical processes and external events in the environment is crucial in creating the slow-motion effect.

Of course, reports by people who have survived dangerous situations are necessarily memories and of a subjective nature. Indeed some doubt the veracity of these accounts, and think that in the process of looking back at the dramatic events a slow-motion effect has been read into their memories and that, for the observer in this situation, the world didn't really slow down. For this reason, psychologists and neuroscientists are attempting to at least approximate this phenomenon in real time in the laboratory.

In psychological research, an approach that is frequently used to record alterations in time perception is the use of time-estimation tasks. In these tasks, test subjects have to judge the temporal duration of emotionally charged stimuli.[9] Emotional stimuli receive particular attention, and the involuntary emotional reactions can be recorded. Thus, if attention is heightened on the one hand and the level of physical arousal increases on the other—both factors modulate the experience of time—the subjective experience of duration should alter as well. Typically, these psychological experiments show that negatively experienced noises (babies crying, a couple arguing) or photos (pictures of gruesome accidents or of rattlesnakes ready to strike), which are presented for just a few seconds and which go hand in hand with measurably increased physiological stimulation of the heart, breathing, and sweating, are temporally overestimated by comparison with more neutral stimuli (the sound of a lawnmower, pictures of household objects).[10] But equally in the case of emotionally positive stimuli, such as erotic photos, delicious food, or atmospheric music, the phenomenon of time expansion can be seen.[11]

Stimuli moving toward an observer are interpreted as potential sources of danger because they are on a "collision course." By comparison with static stimuli or objects moving away from the observer, objects looming virtually toward the observer are overestimated in their temporal duration.[12] While such laboratory tests are a first step toward controlled investigation of the slow-motion effect in danger situations, they do not really represent terrifying situations, which often appear unexpectedly and are actually threatening. Moreover, the effects are comparatively weak. In these experiments, subjects' experience is indeed that an emotional stimulus seems longer than a neutral stimulus even if both are presented in equal duration, but the differences are marginal. The rather boring laboratory setting in which test subjects sit, looking at images and pressing buttons, is far removed from the reality of an actual danger situation. The laboratory tests concern relatively moderate alterations in experience, but they clearly indicate the direction in which subjective time might vary in intensely emotional situations.

A real "danger situation" outside the laboratory was set up in a study carried out by an American researcher, David Eagleman. In an amusement park, test subjects had to plunge in free fall from a 31-meter-high tower into a net, a task that produced a healthy degree of fear in each of the test subjects—after all, the height of the fall was more than three times higher than the 10-meter diving tower in an open-air pool. Nevertheless, the slow-motion effect that was expected to occur during the fall could not be demonstrated. The speed of information processing, heightened by the excitement created during the fall, should have led to an increase in temporal resolution as experienced on a flickering visual display attached to a wristband. If the outside world had actually "moved" in slow motion, the falling test subjects should have been able to perceive a faster rate of flickering of the LED lights than they could see under normal conditions—an effect that did not occur.[13]

Here we can see the disadvantage of field studies that are difficult to control. One question is to what extent the test subjects were actually able to look at their wristband displays during the fall. In this case, experimental research lags behind the phenomena, since, for understandable reasons, real danger situations can only be reproduced to a limited extent. Nevertheless, however, this first attempt at controlled field research could show other scientists who are studying the slow-motion effect how to get out of the laboratory and into the real world. There are in fact people who of their own free will engage in activities that actually put them in life-threatening danger, for example kayaking over huge waterfalls, surfing monster waves, or jumping off cliffs in a wingsuit. There is no shortage of reports of slow-motion effects experienced in extreme situations while pursuing such activities.[14] Even less dangerous pursuits, such as bungee jumping, are practiced in order to bring about particular physical reactions and their concomitant extraordinary states of consciousness. Thus there could be various points of departure for field studies to investigate the slow-motion effect.

One effect was obtained from the aforementioned free-fall study, however: the individuals falling from the tower estimated their own fall as lasting longer than the free fall of other people, which the test subjects observed afterward from below. The researchers working with David Eagleman considered this an indication that people do not experience an alteration in time during the terrifying situation, but rather the feeling of time expansion occurs after the event—that is, the interpretation is related to the memory of the event. What an individual experiences in a moment of terror is so exciting and significant that many more details can be remembered afterward. According to Eagleman, this is what leads to the feeling of time expansion.

In fact the wealth of events in our memory are related to the felt duration of the experiences. A retrospective judgment of time is based on completely different factors from the prospective judgment of time made about the flow of time that is just being experienced, a judgment in which attention and physical sensation are crucial. The more powerfully experiences are remembered, the longer lasting a past period of time seems to us. In retrospect, we make judgments of time when we become aware of an elapsed time span and think, for example, that an entertaining movie went by much too quickly. For past periods of time, memories must be actualized, and the more changes of events are remembered, the longer the past interval of time

seems to us. This applies to shorter as well as to longer periods of time: to the recently seen movie and to a decade of our
life.[15] The emotional situation of a free fall during a period of just
a few seconds seemed in retrospect longer to the test subjects
than another person's fall, which was observed in a relaxed situation from the ground. The memory of one's own fall is full of
emotionally charged thoughts and experiences and accordingly
seems longer.

Both temporal perspectives certainly interact. In moments
of excitement, time is experienced as slowed down, in extreme
danger situations even in slow motion. Through the intensity
of experiences, in retrospect a period of time is remembered as
longer lasting. Memory research tells us that emotional events
are remembered better and with more detailed precision. In this
way, the events expand even in retrospect.[16] This also explains
the results of a study of skydivers undertaking their first jump.[17]
The more fear the novices felt, the longer their first jump lasted
for them subjectively: from boarding the airplane, ascent, and
preparation for the jump to the jump itself. In all, it lasted some
30 minutes. At any given moment, the more anxious novices
experienced their body to a more intensive degree, which led to
an expansion in felt time. We can also imagine, however, that
in retrospect the more anxious skydivers were able to remember
many more emotionally freighted details, and so the entire time
period expanded perceptibly.

A similar effect of time expansion could also be shown in a
laboratory study directed by Olga Pollatos of Ulm University.[18]
In this study, students overestimated the length of a short video
clip from a horror movie by comparison with an amusing and
entertaining movie and a documentary film. The structure of
the experiment was as follows: the students were first shown

the three different video clips—all of 45 seconds duration. They were a funny cartoon, a factual documentary film, and a scary horror movie. The students were told only afterward that they had to estimate the length of time. It was thus the retrospective judgment of time that was being tested. The horror film was remembered as the longest, followed by the documentary and the relaxing cartoon, which was considered the shortest. The fact that the entertaining movie was subjectively the shortest corresponds to the impression that time flies when you're having fun. In a second section of the experiment, other students had to watch the three movies, paying particular attention to their physical reactions. This orientation toward attention led to an even more intense time expansion when watching the horror movie, which elicited the most emotions. This is another powerful indicator that the perception of bodily processes is connected to time consciousness.

Time and Space: The Intoxication of Drugs

"The hashish eater's demands on time and space ... are absolutely regal. Versailles, for one who has taken hashish, is not too large, or eternity too long." This statement on the interrelationship between the experiences of time and space and their massive alteration under the intoxicating effects of hashish was made by Walter Benjamin in Marseille on September 29, 1928.[19] Over the years Benjamin undertook several controlled experiments on himself, at which both the philosopher Ernst Bloch and two doctor friends were present. On this particular evening, Benjamin reported that he took hashish after a long period of hesitation. This time he was alone. As he writes in the record he kept of the experiment, he had been in Aix-en-Provence during the day. He experienced uneasy thoughts; a small child crying next door disturbed him; he lay on the bed smoking a cigarette; reading Hermann Hesse's *Steppenwolf* provided him with a final impetus, and so he ate a piece of hashish. To begin with, Benjamin reports, the drug had no effect. Finally, he left the hotel and went to visit the cafés and bars on the port. On the following pages, which he wrote the next day, Benjamin reports in detail on his altered perceptions as the effect kicked in: "The handle of a coffeepot suddenly looks very large and moreover remains so" (p. 48). "Now ... began the game, which I played for quite a while, of recognizing someone I knew in every new face" (p. 50). Benjamin describes feelings of loneliness and happiness and indulges in artistic and philosophical observations. Moreover, he had an insatiable appetite and ate dinner in two different bars on the port. From this evening episode and the other collected accounts that can be read in the small volume titled *On Hashish*, we can easily find clues explaining which factors play a role in the alteration of time consciousness during hashish intoxication.

Aldous Huxley, Charles Baudelaire, Ernst Jünger, Pitigrilli, and Jörg Fauser also described altered states of consciousness under the influence of drugs such as alcohol, opium, mescaline, and LSD. In addition, scientifically minded researchers such as Alexander Shulgin, John Lilly, and Albert Hofmann have reported in detail on the experiments carried out on themselves. We take these substances primarily because of the effect that intoxicates and makes us feel happy, and sometimes also because we want to distract ourselves from a life that we experience as boring. The intoxication offers the opportunity of forgetting the self, or rather the everyday self. We attain a dissolution of the self, as reality disappears and the boundaries between subject and object are blurred. The events of the world combine associatively to form a meaningful whole, revealing new and surprising perspectives.[20] As can be seen from Walter Benjamin's account, he too is fleeing a situation of lonely restlessness, and experiences meaning and purpose under the influence of hashish. In this case, if time expands, then it is into a state experienced by the individual as generally pleasant and extremely interesting, comprising all one's thinking and feeling.

Many anecdotal accounts of the powerful expansion of time when consuming hashish can be understood as descriptions of an initially more powerful sensory sensibility, intensity, and density of experience that has an impact on memory. "The memory of the intoxication is surprisingly clear" (p. 118); "colors grow brighter, more luminous; objects more beautiful, or else lumpy and threatening" (p. 117), as Benjamin writes. In this the experiences are intensive enough to create the feeling of time stretching, despite the proven partial impairment of the memory function under the influence of hashish when looking back at the experience. Short-term memory is particularly affected by acute cannabis consumption, for example when the beginning of a sentence, having been started, is no longer present at the end, and the statement tails off, as the content and intention of what was meant has been lost. Conversations stop because the conversation partners suddenly no longer know what they were just talking about. The memory processes of storage, retention, manipulation, and recall are demonstrably impaired in the case of intense, chronic cannabis consumption that has been practiced over many years.[21] In an acute state, however, the experiences are so vivid and often so bizarre that they remain more strongly in the memory despite the memory lapses. "In the night, the trance sets itself off from everyday experience with fine, prismatic edges. It forms a kind of figure and is more than usually memorable," as Benjamin puts it (p. 53). Here too, in retrospect, the recallable memories create the perception of duration: under the influence of psychoactive substances, intensively experienced moments appear in hindsight to last longer because a larger number of intensive experiences are remembered.

But in the lived moment too, under the intoxication of hashish, time expands, particularly during a period of waiting: "Then it began to take half an eternity until the waiter reappeared. Or, rather, I could not wait for him to appear" (p. 51). Benjamin's friend, the doctor Ernst Joel, describes the torturous experience of waiting: "Along with this, considerable miscalculation of the time—so impressive that at one point I thought my watch was running backward" (p. 40). The felt duration of time expanded to such an extent that the actual duration of time according to the watch was unbelievably short. As we have seen, in moments of waiting our attention is focused more intensively on the time, which leads to an overestimation of duration. Laboratory tests conducted under strict experimental conditions, using volunteers to look at the perception of time under the influence of cannabis, were able to substantiate the relative overestimation of time duration, in line with the accounts provided by Benjamin and others.[22] A double-blind, placebo-controlled study using various administered doses of cannabis showed a clear overestimation of temporal duration, ranging from seconds to minutes.

Alcohol, by contrast, makes time pass more quickly. Even moderate quantities of alcohol impede the storage of events and lead—when regarded retrospectively—to periods of time that are experienced instead as shortened. This has been demonstrated by controlled tests on the perception of time under the influence of alcohol.[23] The time spent at boring parties or at events where one feels alone can be reduced by alcohol, at least subjectively. It is thus not by chance that alcohol is the friend of the lonely and brings people together. As Charles Baudelaire wrote: "Wine knows how to embellish the most sordid room / With a luxurious disguise."[24] Despite an intensified sense of perception, not dissimilar to the effect of cannabis consumption, alcohol reduces the ability to concentrate and thus creates the experience of time passing more quickly.

Anecdotal reports of time "racing" occur after the ingestion of cocaine or Ecstasy (MDMA)—for example, being greatly surprised that one has danced the night away and the sun is already coming up. Cocaine and Ecstasy are stimulants that produce a feeling of euphoria and help overcome social inhibitions, which is why they are often used as party drugs. Sigmund Freud's letter to his future wife, Martha Bernays, in which he describes in euphoric terms his experiments with medical cocaine, is famous:

> Woe to you, my Princess, when I come. I will kiss you quite red and feed you till you are plump. And if you are forward you shall see who is the stronger, a gentle little girl who doesn't eat enough or a big wild man who has cocaine in his body. In my last severe depression I took coca again and a small dose lifted me to the heights in a wonderful fashion.[25]

Studies show a relatively high speed of thought processes after amphetamine derivatives such as D-amphetamine (dextroamphetamine) have been administered.[26] In laboratory-controlled tasks involving the perception of time, the accelerated processing leads to the normal overestimation of time intervals in the range of seconds. In the case of longer time intervals, such as hours, however, extremely accelerated thinking and perception mean that experiences are not stored in a sustainable way. One has scarcely had one thought before one is already onto the next thought and the next perception. Switching one's attention at high speed leads in hindsight to less pronounced memory traces—with the effect that the hours experienced are seen in retrospect as shorter, and time passes more quickly. This is comparable with the feeling of a hectic day spent in the office, where one engages in lots of activities, switching quickly between them without going deeply into things. If one looks back at the end of the day, the hours passed extremely quickly, and one wonders what one has achieved today.[27] Quite a lot, actually, but nothing given one's full attention. For this reason fewer memory traces are left behind and the working day has passed quickly.

Caffeine too is a stimulant, as the majority of the coffee- or tea-drinking global population knows when they want to "get going" in the morning. Anyone who has ever had an overdose of caffeine can perhaps imagine how illegal stimulants with the risk of addiction have an effect on the speed of thought and temporal experience. Thoughts race and actions speed up. But everything happens so quickly that experiences do not embed properly. Like other stimulants, an overdose of caffeine engenders constant mental leaps and a lack of deep processing; longer periods of time pass more quickly as ultimately fewer memories are stored.

Benjamin's observation about hashish consumption, with which we began this section, illustrates the change in the perception of time and space caused by cannabis; both are suddenly experienced as disproportionately large. We might say that in the process of experience, time and space were interlinked, as their dimensions expanded together.

In classical physics—but also in the theory of relativity—time and space are inseparably linked. Movement can only be described as a change in space and time. A near-collision of two vehicles, for example, can take place in terms of spatial data (centimeters) or temporal data (milliseconds). Ultimately, at the macroscopic level, physical time can be defined exclusively by space, as movement in space;[28] that means that the determination of time of day takes place as a change in space. For example, units of time are defined as the number of swings of a pendulum, or—put more generally—as periodic events. Such periodic events define the day and the year, brought about by the earth turning on its axis and orbiting the sun. Duration is visible in smaller units of time as the progress of a hand on a clock face, or in the case of ancient clocks, as a particular quantity of water filling a vessel.

The mental representation of time is also linked to the spatial one.[29] For example, we associate earlier times with the left-hand side of a sheet of paper (according to René Descartes's coordinate system where the x-axis can represent the arrow of time) and later times with the right-hand side; for us, time proceeds from left to right—that is, in space.

Time seems "long" to me; I use spatial metaphors to describe the experience of time. When looking at a day or a year, time can also appear circular; but this is another spatial conception. Moreover, in most cultures the past is associated with what is located behind us, while the future is what is located ahead of us. In this notion, the individual's own body creates the relationship to the present. Memory and anticipation are spatially oriented according to my physical position; I have left the past behind me, and the future lies before me.[30] Typically, we represent time using a mental line on which "before" and "after" are positioned on the left and right relative to one another. In this way time is illustrated through space.[31] When we cover distances that require a certain amount of time, it becomes clear how the representation of time and space are mutually determined. Gaining time by speeding up our means of transport also means that space

Figure 1.1
My jogging route in La Jolla, California: Black's Beach. From the runner's
perspective there is both an extent of space and a period of time to cover.
The space of time is indissolubly space and time.

is swallowed up. A journey that is experienced as lasting longer expands the experience of space. And conversely, the more
I experience on a journey, the more I see of the route and the
longer the trip lasts (see figure 1.1).[32]

Neurology has given us clues to a kind of mental organization
that underlies both time and space, and which is combined with
physical sensation. For example, patients with particular dam-
age to the right side of the brain who suffer from the disorder of
visual neglect ignore their left side. In extreme cases they eat only
from the right-hand side of the plate and dress only the right
half of their body. The left-hand side of the world is perceived

inadequately, and attention is directed only to the right-hand side of the world. As was shown in a study, these patients also ignore events from the past, which are of course located on the left-hand side of the mental arrow of time.[33]

These reports provide further proof that the representation of time is linked to a functioning perception of space and body. Accordingly, Walter Benjamin's hashish intoxication altered neurophysiological processes that underpin the perception of time and space.

For the French philosopher Henri Bergson (1859–1941), immediately experienced duration (*la durée*) as conscious experiential time was in the first instance an "intensive," nonquantifiable factor. Time consciousness and self-consciousness are combined, because states of consciousness—more or less intensively—develop in time and thus constitute duration.[34] Subjective time can be quantified, that is, spatialized, only by comparison with external factors, such as clocks for measuring objective time or natural changes such as the transition from day to night. Time is understood via a spatial factor. Only because of this are we able to say that the pain was just a few seconds "long." We imagine a linear arrow of time onto which the beginning and end of the pain can be mapped.

This "double aspect" of subjective time (as Bergson formulated it) can also be found in the two-step model that forms the neuronal basis of temporal experience as described by the French neuroscientist Virginie van Wassenhove.[35] In an early stage in the process, temporal information is implicitly present in the dynamic of the neuronal processes. In this way the spatio-temporal operations of the brain are associated with conscious experiences of color, pitch, or an emotion. However these neu-rophysiological processes don't entail a ticking clock to measure time explicitly. The neuronal processes of perception become the experience of duration via a second step. This second stage in the process involves the depiction of time in cognitive metaphors of space, using knowledge of clock time. Pure experience of temporality becomes perceived time in space.

Most accounts of the effects of drugs include reported perceptions of space and time that are, at least in part, highly distorted. To a great extent these effects can be explained within the framework of temporal perception's classic models of distracted attention, stored memories, and changes in physical state. However, Benjamin is already using terms such as "eternity" and "immeasurable space," where the conventional understanding of time and space is stretched to its limits. (See figure 1.2.) In the case of hallucinogens such as LSD and psilocybin in particular, experiences occur in which time and space actually seem to dissolve (more on this later). In opium intoxication, too, the sense of time and space can alter so massively that it leaps from a still comprehensible disproportionality to a size that is no longer graspable, as Thomas De Quincey powerfully described in his 1822 *Confessions of an English Opium-Eater*.[36]

In his description of the intoxication, which was not based on external perception but rather replicated his inner experience, that is his trip,

> the sense of space, and in the end, the sense of time, were both powerfully affected. Buildings, landscapes, &c. were exhibited in proportions so vast as the bodily eye is not fitted to conceive. Space swelled, and was amplified to an extent of unutterable infinity. This, however, did not disturb me so much as the vast expansion of time; I sometimes seemed to have lived for 70 or 100 years in one night; nay, sometimes had feelings representative of a millennium passed in that time, or, however, of a duration far beyond the limits of any human experience.

Figure 1.2
Specimens of the long-lived bristlecone pine (*Pinus longaeva*) in the White Mountains in California, where the oldest tree is over 5,000 years old. Although such long periods of time cannot be experienced directly by humans, there are nevertheless reports of drug-induced dreams and time travels in which the subject lives for thousands of years. Perhaps these experiences are due to emotions that are evoked by chronologically narrated stories, for example, imagining that the trees in this photo were already a thousand years old at the time of Socrates and Jesus.

But what remains of the euphoria of intoxication? Some people integrate their dealings with consciousness-altering substances as an important experience in their lives, in particular in the case of hallucinogens that do not cause physical dependency. However, even these substances are not without their dangers, as they can induce psychotic episodes in those of a susceptible disposition. On the other hand, heroin or stimulants such as cocaine and methamphetamine harbor the danger of addiction. Even Sigmund Freud must have realized this after his initial euphoric declarations about the potential of cocaine: he later spoke of cocaine as humanity's third scourge, after alcohol and morphine. The condition of substance addiction, moreover, also leads to an altered perception of time. This is not in any way comparable with the extraordinary time consciousness of intoxication, but becomes evident in more mundane and unpleasant ways. After an extended period of consumption, abstinence triggers physical withdrawal symptoms that disappear only with resumed use of the drug. Indications of physical dependency—regardless of whether upon nicotine, alcohol, or illegal substances—are repeated compulsive consumption despite the intention to abstain, and an increase in the dosage required to attain the same positive effect once again.

In withdrawal, however, time expands. Physical symptoms become noticeable—restlessness, shivering, an intense craving for the drug. Thoughts revolve solely around the object of desire. It's not necessary to imagine heroin withdrawal, as smokers and former smokers know this condition too. One lives unhappily in the present, and one's relationship to the present is heightened— impulsively related to the substance. Future events are very far away. This means that decisions in favor of possibly sensible but long-term options are made more rarely; short-term, but perhaps less advantageous options are favored instead.[37] This is the very crux of addiction: I would rather smoke a cigarette now than resist temptation now and profit from this in the long term. Periods of time stretching into the future seem very long. To the addict, it seems impossible to survive this time without recourse to the addictive substance. In a study it was indeed shown that abstinent smokers with a serious need for a cigarette estimated time as passing slower than smokers with a less pronounced desire.[38] In another study, which I carried out with Martin Paulus at the Veterans Affairs Medical Center in San Diego, patients in the drug treatment clinic, who had a long-standing dependency on cocaine or methamphetamine, estimated a 53-second interval as lasting 24 seconds longer than the control subjects did.[39] This is empirical proof that in drug addiction the perception of time is actually altered. The findings also show that addicts estimate periods of weeks and months as seeming longer than other people do.[40] The intensified relationship with the present and the perceived time expansion in the expectation of future events make it hard to imagine oneself in a drug-free future. The present is overly present. In a state of addiction the individual loses his or her temporal freedom—the freedom to choose between present and future opportunities.

At One with the World

As is clear from case study reports, alterations in time consciousness are a typical indication of an extraordinary state of consciousness. The feeling of time expansion and the deceleration of events is the extreme manifestation of the usual experience of a linear progression of time. Time is still directional and proceeds in the flow of expectation, experience, and memory. Events are initially anticipated as in the future, then they are lived in the moment, until finally they become the past as memories. In states of consciousness of danger such as those already discussed, the time of the seconds as they are lived still flowed in linear fashion: the truck drives toward the car; the driver's quick reaction prevents the collision; the truck manages to miss the car (by a matter of centimeters or split seconds). The events proceed in a chronological sequence. Equally, under the influence of drugs, as we have seen, an alteration of linear, "spatialized" time occurs: periods of time expand. However, time and space are still experienced in humanly understandable dimensions. Nevertheless, Thomas De Quincey's opium use led to an experience of time and space that he described as "far beyond the limits of any human experience."

There is an extensive body of accounts concerning phenomena in which time is altered in extraordinary states of consciousness, as can occur under the influence of drugs, but also in meditative contemplation, in rhythm-induced trance, or in the extreme case of near-death experiences.[41] Felt time becomes slower and slower (but sometimes also faster) until it reaches a standstill. The perception of linear time transforms into timelessness.

Accounts of mystical states arising, for example, in meditative contemplation or through spontaneous experience, describe the disappearance of linear time. They result in the experience of time standing still, or even the loss of the sense of time.[42] Perceived moments expand to such a length that future, present, and past become a felt unity of "eternal" present. These states of consciousness of timelessness and eternity—as the copresence of past and future events—often go hand in hand with the removal of physical and spatial boundaries and a feeling of happiness in being at one with the universe. This kind of experience is reflected in theologically inspired descriptions of God provided by late medieval mystics such as Meister Eckhart:

> In eternity there is no before and after, and what happened a thousand years ago and what will happen in another thousand years is one in eternity. Therefore what God did, what he created a thousand years ago, what he will do in a thousand years' time and what he is doing now are all one.[43]

What is mysticism? The philosopher Ernst Tugendhat makes an interesting distinction, seeing mysticism and religions as two separate ways in which people deal with the experience of suffering and misfortune.[44] According to this, religion provides a wish projection related to one's own self (for example living in paradise after death).[45] Divine power or the gods determine personal happiness, and they can be influenced by ritual, magic, or prayer. According to this explanation, for the religious person it is primarily about him- or herself and the hope for improvement in their own state of affairs. By contrast with the religious person's "egocentricity," mysticism (or spirituality) in essence entails either denying oneself and one's will through complete renunciation or at least relativizing it. The aim of the mystic, Tugendhat concludes, is peace of mind, which, as a liberation from suffering, is attained through the retreat from desire and greed. Fundamentally, this is about calmness in the face of the contingency of the world, whose events can only be controlled to a limited extent and all too frequently mean misfortune and hardship. In such a state of calm self-abandonment, the spiritual person liberates himself from what binds him, his self-love, his will, his ideas, his physicality, and his temporality—from the memories that never let go, the demands of the present, and plans for fame and fortune in the future.[46]

Although there are many different kinds of and approaches to the spiritual experience, it is always about the merging of the individual with the whole, about sinking into the encompassing, the experience of the absolute, of transcendence. In the Christian context it is about knowing God as becoming one with God, the *unio mystica*.[47] In mysticism, it is in principle about the sublation of the subject-object dichotomy or, as Karl Jaspers put it, about the total "union of subject and object, in which all objectness vanishes and the I is extinguished. Then authentic being opens up to us, leaving behind it as we awaken from our trance a consciousness of profound and inexhaustible meaning."[48]

In the spiritual experience, self (subject) and world (object) are no longer separate. Here we can differentiate between two types of being at one: the self merges with the encompassing (multiplicity dissolves; everything becomes one); or, on the other hand, all objects in the world are related (everything is connected to everything else). After encountering the whole (the divine), the individual experiencing this has a feeling of inner peace and empathy with others.

The transition from linear to mystical time is a universal experience in all spiritual traditions and world religions, engendered by the techniques of prayer and meditative contemplation.[49] Crucial to the disappearance of temporal and spatial intuition is the dissolution of the self in becoming one with the encompassing. Temporality is connected to the conscious experience of my self in space and in time; if I am conscious of my self, then I feel the progression of my physical and mental self over time. If dissolution of the self occurs, then the understanding of time and space dissolves as well. An inherent consequence of egolessness is timelessness, as well as a lack of sense of body and space.

According to Immanuel Kant, time and space are forms of pure intuition of the transcendental self. Time and space are not conceivable without the self. Equally, in phenomenological observations on the experience of time, for example, consciousness of self and consciousness of time are considered interdependent. I can only become conscious of my self if I experience my momentarily perceived self projected into the past (remembered self) and the future (anticipated self).[50] While change occurs all around me all the time, the conscious perception of my self persists in time. Without a concept of self, time does not exist.

Mystical experiences are not restricted to the masters of spiritual traditions. Even people who have no intellectual or spiritual background can have them.[51] Examples are the moments filled with intensity and clarity that suddenly come upon people, sometimes after periods of extreme concentration and exertion, but which can also happen spontaneously while one takes a walk in nature. Extraordinary experiences of extreme intensity and great happiness—magic moments that are remembered for a lifetime—are felt, for example, by a child at Christmas as the harmony of festive family togetherness. Lovers experience them in the first weeks of their ecstatic, physical "oneness," and in a different way after decades of intimate cohabitation and understanding. However, they are also available to attendees at a party, who unexpectedly experience a feeling of close community; or to a sports team, which has trained for months toward a common goal that has now been attained.

The feelings that arise in these circumstances can scarcely be put into words: they must be experienced. They are most likely to be translated into music, as music can evoke emotions. An individual listening in concentration to his or her favorite music (or who is amazed by a new piece) can be absorbed by it, losing themselves intellectually and emotionally, and attain a trancelike state in which even the concepts of space and time are altered.[52] When we play music or dance, music becomes an emotional and physical experience that alters consciousness and brings people together in a common experience.

The longing to become one in love and the impossibility of fulfilling this in life was realized textually and musically by

Richard Wagner in his music drama, *Tristan und Isolde*, as the musicologist Egon Voss explains:

> His [Tristan's] longing is insatiable, as it is nothing less than the desire to sublate individuation, to overcome the boundaries of the self. ... The love scene in Act Two ... is obviously nothing less than the merging of I and You, and their transcendence into the entirety of the cosmos. Love can also satiate this longing. But the fulfillment it provides does not last.[53]

Isolde expresses this, as she bends over Tristan's dead body and sings the last words of the opera, beginning with "How soft and gently":

> In the heaving swell,
> in the resounding echoes,
> in the universal stream
> of the world-breath—
> to drown,
> to founder—
> unconscious—
> utmost rapture!

The contingency of life, the general randomness of life's circumstances, too often prevents the fulfillment of love. By contrast with our daydreams and the products of the entertainment industry's dream factories, in which a "happy ending" can be imagined into existence almost effortlessly, all too often reality brings only disappointment. We learn that we are dwelling in reality and not in a daydream, as our wishes are not fulfilled in the twinkling of an eye, but instead encounter resistance.[54] Tristan and Isolde have already met, but because of complicated circumstances it is impossible for them to be together. Now they encounter each other once more, but Tristan has to take Isolde to be married to King Marke, to whom she is betrothed. Once again, circumstances stand in the way of love. Merging into love fails, and all that remains is the merging of the self into the whole through one's own death, the *Liebestod*, or love-death. In an ideal situation, if luck would have it, love can at least temporarily satiate the longing for the merging of the self with the other into a greater whole. However, an extreme desire is also the source of disappointment.[55] For this reason teachers in the spiritual traditions talk about relativizing desire related to self, in order to avoid disappointment and pain. In the dissolution of self they are aiming at, there is neither desiring subject nor object of desire.

Alas, few are granted a life-changing, mystical experience completely spontaneously. Incorporated into a spiritual context, however, meditative practice can be a way of experiencing the encompassing. Essentially it is a question of techniques for managing the experience of time. As Wolfgang Achtner, a theologian at the University of Gießen, has shown convincingly in an essay, across cultures and religions in general it is a question of deepening the experience of now.[56] The technique recommended by medieval Christian theologians and Buddhist teachers of meditation, independent of one another, is concentration on the present moment. The experience of self is dependent on consciousness of the past and the future. Accordingly, it can be reduced to the experience of now through the method of focusing. By practicing concentration on the moment one ultimately reaches the state of timeless eternity. According to Meister Eckhart, the practitioner must first apply the entire will to achieving timelessness beyond the experience of now; subsequently, however, the activity of the will is dissolved.

In meditation, processes of anticipation and memory are closed down. Only the perception of the moment is allowed:

now the breath, here the perception of the body (we will look into this in more detail later). What can happen in meditative contemplation, after persistent practice, can be described as follows:

> During contemplation the stream of experience slows down more and more. ... The individual in contemplation experiences the self in a way that is no longer moving forward but motionless. He speaks of "motionless calm." Calm is motionless in him and he is motionless in calm. One might think that time seems to stand still. Even the individual in contemplation knows, always in the sense of co-consciousness, that time is and that he is in time, but his experience in the persistent immutability of calm is as if time stands still.[57]

Timeless in Near-Death

Timelessness can, however, be experienced in two ways:[58] (1) It can manifest itself in the sense just described, initially as the perception of an alteration in the speed of time passing that leads in extreme cases to time standing still or the feeling of eternity. (2) But timelessness can also be expressed by time losing its meaning; temporal processes and the surrounding space are sometimes clearly perceived, but time and space have lost any kind of relevance for the individual affected. This second form is reported by people who have experiences using ayahuasca,[59] a psychoactive substance consumed in South America in religious and ceremonial contexts, or with mescaline, for example by Aldous Huxley:

> Space was still there; but it had lost its predominance. The mind was primarily concerned, not with measures and locations, but with being and meaning. And along with indifference to space there went an even more complete indifference to time. "There seems to be plenty

of it," was all I would answer. ... Plenty of it, but exactly how much was entirely irrelevant.[60]

As perhaps the most extreme human form of altered experience, near-death experiences too can produce the feeling of timelessness. In *Own Death*,[61] the Hungarian writer Péter Nádas writes about a heart attack and the resulting insufficient supply of blood and oxygen to the brain. When he wakes up in the morning he feels different, but goes into town anyway. Something is happening to him, his body won't work properly, he avoids the scorching sun, he's hot, and while being treated at the dentist's he's drenched in sweat, but: "I didn't understand what was going on" (p. 17). Even the exhaustion, which means he can scarcely move forward while climbing a hill in the afternoon, he interprets as weakness because of hunger. Then vague pains appear, alongside fear, but then he feels better and carries on. Insidiously at first, the symptoms increase more and more markedly. He sits in a pub and lights a cigarette, which makes him breathless. From time to time he has lapses in consciousness: "then nothingness, absolute nothingness" (p. 45). He carries on sitting at the table but doesn't touch his soup. The fear of death grips him, a throbbing, stabbing pain in his right shoulder. His breathing difficulties force him into the restroom, where he washes his face in cold water. He no longer recognizes himself in the mirror, but he sees that all the color has drained from his face. He goes back to the table and signals to the waiter, and finally leaves the pub. He is constantly short of breath, and this doesn't change even outdoors. He returns to his apartment in a taxi. The shortness of breath remains, and his chest pains suddenly increase. During this whole time, Nádas doesn't think about sending for help; rather his thoughts are focused on the proof corrections of his forthcoming book. But it gradually becomes clear to him that it must be a heart attack and that an afternoon nap on the sofa, on which he is now resting, won't bring the necessary help. However, he must have fallen asleep at some point, as the clock has suddenly moved on to seven o'clock in the evening. Now he does set off for the local doctor, whose surgery is still open on the other side of the street. But he can hardly walk, and leans against the walls of the houses. The people around him behave as if they didn't see him.

Then something begins "that is very hard to grasp with words," as Nádas writes, "because in this pre-death state, conventional chronology loses its meaning almost completely. A big switch is turned off, the main switch. Whereupon seeing, perceiving and thinking by no means stop. However, these simultaneously operating functions don't string their freshly acquired impressions on the conventional timetable of the mind" (pp. 121–123). The near-death experience.

> Timelessness reigns in the universe. One might call it a cosmic experience. ... It couldn't in good faith be called a space. I saw my past life through a medium that, with its temporal order, took its place in the immeasurable void of timelessness. ... The first and the last time are not separable. (p. 213)

> The mind deprived of its bodily sensations perceives the mechanism of thinking as its last object. (p. 129)

Time, space, and physical sensation have dissolved. He is experiencing a state of extreme ecstasy, such as he has never physically felt before. He cannot articulate this sensation verbally. He is conscious that he is now dying. At the same time, he watches how the paramedics tend to him and take him to the hospital. Because space and time do not exist, he has an overview of his entire life in its totality. In this, something happens that Nádas calls a "flipping or tipping move" (p. 209), a dazzling light penetrates his consciousness, and a force carries him toward the distant source of light. It is, "as the world appears to one seen from the bottom of a cave on a cloudy day" (p. 225). When he watches the paramedics breaking his ribs, he is again present and is watching what is happening from a somewhat elevated position. The force again draws him toward the light, but he doesn't reach it. The light source disappears and he finds himself once again under the neon lights of the hospital room, where people are bending over him. They notice that the patient is coming round, is alive, and they burst out laughing with joy: "Which meant that I was back again" (p. 245).

The Dutch cardiologist Pim van Lommel has collected and analyzed hundreds of near-death experiences of people who have suffered a cardiac arrest that lasted minutes.[62] Because of insufficient blood and oxygen supply to the brain caused by circulation failure or respiratory standstill, people lose consciousness. They are considered clinically dead, as the brain has lost its ability to function and will be irreparably damaged within minutes if they are not resuscitated. Nevertheless, after the event, some people report extraordinary experiences of particular intensity during the time they were unconscious. According to estimates, between 10 and 20 percent of people who were clinically dead and then resuscitated report such an event.[63] The experiences are characterized by a feeling of calm and peace, and sometimes go hand in hand with the feeling of slipping out of the body. Accounts describe a powerful light at the end of a tunnel, and sometimes a meeting with deceased relatives in another world occurs. These experiences can be strongly character- and life-changing. Van Lommel summarizes the near-death experiences (NDEs) thus: "People are transformed by the glimpse of a dimension where time and space play no role, where past and future can be viewed, where they feel complete and healed, and where infinite wisdom and unconditional love can be experienced" (p. 329).

The aspects of near-death experience that van Lommel maps out highlight powerful similarities with other extraordinary states of consciousness, such as appear in mystical experiences or can arise through the administration of hallucinogenic drugs. These are borderline experiences of particular intensity that place people in another relationship to their life and individually redefine their life's meaning. Anyone with such an experience often fears death less afterward and lives more consciously and more spiritually. The numerous experiences of people that have been collected highlight powerful similarities, but there are clearly social and cultural specifics as well. This was shown, for example, in a study carried out at the end of the 1990s, in which near-death experiences of East Germans and West Germans were compared who for decades had been separated by the iron curtain.[64] Classic symptoms of a near-death experience, such as seeing a light, as well as the feeling of floating outside one's body and being in another world, were more pronounced in West Germany. In East Germany, by contrast, the experience of a tunnel was more frequently reported, and the NDEs more often had negative connotations. Since near-death experiences had been frequently discussed in the media in West Germany since the 1970s, and books on the subject had become bestsellers, but these experiences were scarcely addressed in the GDR, this suggests that previous sociocultural experience influences the characteristics of near-death experience. In addition, in East Germany there is a substantially greater percentage of self-confessed atheists; spiritual and religious influence, too, is a contributing factor to the experiences reported. Nevertheless, it is still true that there are recurring phenomena that are reflected in the realms of experience of children (who had no knowledge of near-death experiences) and adults of all cultures, and even eras, since written records began.[65]

Research into near-death experiences shows a way forward to finding out more about the great mystery of consciousness. Anyone who reads without preconceptions the accounts of patients, in conjunction with the clinical details and the reports of medical personnel, will come to the conclusion that the phenomenon is not really understood. Ultimately, there exists an incongruity between the reports of the experience and the clinical reality: while the individuals are seemingly having their experiences, in many cases the oxygen supply to the brain is interrupted because of a collapse in the vital, physical systems of breathing and circulation. For at least a few seconds, brain activity is suspended and the result is unconsciousness. Some patients who are resuscitated and have had a so-called near-death experience subsequently tell of procedures that have taken place during the period of unconsciousness. Sometimes they even report events taking place during the resuscitation attempts, which they experience intensively although they are unconscious. They report details of procedures and conversations that are sometimes confirmed by the medical staff.

The possibility of experiences and memories occurring during unconsciousness that has been caused by a lack of brain activity goes against the materialistic conception of the scientific world. This conception proceeds from the assumption that all conscious experience depends on specifically organized brain activity. If the brain is damaged, for example, this has effects on experience and behavior. Injury to the left side of the brain leads to the impairment of language and speech with substantially greater probability than an injury to the right side. Some mental functions can be localized quite precisely in the brain. Sleeping and waking are linked to certain rhythmic brain signals (fluctuations in brain activity). Caffeine activates the brain and makes us more alert. Other, stronger drugs also have an effect on experience via certain neurotransmitter systems in the brain. Opium, cocaine, and hashish produce such a massive effect because neurotransmitters produced naturally in the body, with the same chemical properties but in much smaller doses, carry out functions in the brain. The substances that come from the outside now flood the brain systems and lead to massive alterations in consciousness. This means that our experience is demonstrably determined by changes in brain activity. If there is no brain activity going on, then it is not possible to experience anything.

This is the textbook explanation. In this context, reports of particularly intense experiences taking place while brain activity has stopped naturally encounter a great deal of skepticism and a huge amount of criticism. For this reason people have attempted to find various explanatory approaches that are compatible with the conception of the world espoused by the scientific mainstream. Ultimately, the person can recount their experience only after they have woken up, and they are thus passing judgment retrospectively. When exactly they had these experiences is not verified. Or perhaps the person in question had their experience while waking up, that is in the period of time when the brain was being resupplied with blood and oxygen in the course of resuscitation, while from the outside it seemed that they were still unconscious? Or perhaps a residue of brain activity remained, which is not so easily verifiable?

We all know that when we are dreaming, real time and dream time do not coincide. Dream experiences that in reality might have lasted only seconds expand in the dream to many minutes, even hours. Is something similar going on during the near-death experience, in a moment when the brain is barely or once again active? It is at the very least astonishing that a similar program of neuronal activity takes place in people of all ages, eras, and cultures, a program that generates this kind of phenomenon in the case of near-death experience.

Other scientific explanatory approaches emphasize the gradual failure of particular regions of the brain. For example, the vision of the tunnel is said to arise because of the decrease in brain activity in the outer visual cortex. Only the regions of the brain associated with the central areas of sight would continue to function in the first instance, leading to the tunnel effect. Overall, according to this view, the phenomena of consciousness in near-death experience correspond with a gradual regression in brain functions.[66] Against this notion of a natural process of deterioration, however, one might argue that the experiences themselves are of such heightened intensity that they can hardly be reconciled with any kind of deterioration. Equally, the tunnel effect of near-death experience is not comparable with the tunnel effect experienced, for example, when one stands up too quickly and becomes dizzy. The experiences are of an extreme intensity and precise detail, so it seems more likely that a particularly efficient brain function is being described. In addition, countering the idea of a process of brain deterioration is the fact that near-death experiences also occur when there is no lack of supply to the brain, for example in life-threatening situations such as mountain accidents, or in warfare, as well as in situations of extreme isolation.[67] There are so many different situations in which near-death experiences occur that the monocausal explanations that are often given do not suffice.

One individual case, which was clinically well documented, demonstrates how poorly understood near-death experiences are. In 1991 Pam Reynolds, an American woman, had to undergo a life-threatening operation. The patient had been diagnosed with an aneurysm in an artery of the brain stem (a dilation of the blood vessel) that could have ruptured at any time, leading to her immediate death. The only possibility was to remove the aneurysm in a challenging operation, during which the blood had to be drained from her hypothermic brain. Thus the patient had to be clinically dead for the duration of the operation, so that the neurosurgeon, Robert Spetzler of the Barrow Neurological Institute in Phoenix, Arizona, could remove the aneurysm from the bloodless brain. During this highly dangerous procedure, the brain was starved of blood for about one hour, and thus completely lacking in neuronal activity. The cooling of the brain made sure that the brain cells did not die. On waking, Pam Reynolds, who survived the successful intervention, reported a classic near-death experience. She even recounted procedures that took place during the operation and gave detailed descriptions of the instruments used, as well as of the team's conversations during the operation.[68] The debate triggered by this case of course turned on the question of when exactly the near-death experience occurred. Was it during clinical death, that is, in a phase when no brain activity whatsoever could verifiably occur? Such a finding would have what it takes to upset the neuroscientific worldview. It would mean that conscious experience may be possible without a basis of brain activity. Or, did the near-death experience take place beforehand, when brain activity was still verifiable but the patient was already under sedation (although that too would be astonishing)?[69]

Perhaps research into near-death experience is a little like the history of alchemy, which, until the early modern period, attempted to find the formula for producing gold. At some point in this process, the formula for making porcelain was discovered. In future research too, evidence will probably not emerge that conscious experiences can take place in an individual without brain activity. One thing is clear, however: in doing this we will discover astonishing things about the brain's capacity to function, as well as about the origin of and conditions for sustaining human consciousness.

2 The Moment

Absolutely in the Moment

Experience takes place in the present moment only. The moment is the immediate connection to reality; I always have inner and external experiences only now.

> The moment is the only reality, the essential reality in intellectual life. The lived moment is the last, blood-warm, immediate, living, the present incarnate, the totality of the real, the only concrete thing. Instead of losing themselves in the past and future, away from the present, the individual finds existence and the absolute only in the moment. Past and future are dark, uncertain abysses, they are endless time, while the moment can be the sublation of time, the present of the eternal.

Karl Jaspers wrote this in 1919 in his *Psychologie der Weltanschauungen* (Psychology of worldviews).[1] In fact, the moment might be as powerful as described here. But, as Jaspers continues, the moment is only too often used as a mere tool for attaining a future goal. Then the moment disappears among the goal-oriented business of the everyday or as the mind wanders into the sometime and someplace. But the self disappears here, too, because only in becoming conscious of my self in the relevant moment can I experience myself unmediated.

And with this we are already immersed in the contemporary discussion and cultural criticism of the industrialized individual, who, in their future-oriented relationship with clock-watching appointments, is said to have forgotten how to experience the present and thus forces the felt acceleration of their life.[2] As the philosopher Byung-Chul Han writes:

> If the goal is the sole point of orientation, then the spatial interval to be crossed before reaching it is simply an obstacle to be overcome as quickly as possible. Pure orientation towards the goal deprives the in-between space of all meaning, emptying it to become a corridor without any value of its own. Acceleration is the attempt to make the temporal interval that is needed for bridging the spatial interval disappear altogether.[3]

In the dominant, functional orientation toward a goal, the present loses its value. We no longer live intensively in the moment, and so life as a whole is lost. Life, both now and in memory, is made up of consciously lived moments. I am present in the moment, meaning I live consciously and intensively, if I give my attention to lived experience. But in retrospect too, life as a whole expands, since it is then full of memories of lived moments. The full life in each felt moment also expands the time intervals in hindsight.

In Jaspers's words, the point is also that "the moment can be the sublation of time, the present of the eternal." This is the mystical conception of the moment as the sublation of time in the dissolution of the self into the whole, into the "surrounding world."[4] This idea can also be found in such an approach to life as formulated by Ludwig Wittgenstein in his famous proposition 6.4311 of the *Tractatus Logico-Philosophicus*: "If we take eternity to mean not infinite temporal duration but timelessness, then eternal life belongs to those who live in the present."[5] This expresses an attitude that is oriented not toward the prospect of life in the future but toward the present in the experience of now. If only we would concentrate on the experience of the present alone, then there would be no death. Ultimately, death is a future event that one cannot experience oneself, since (to echo Epicurus): "While we exist, death is not present, and when death is present we no longer exist."[6]

Of course this is an attitude that cannot be sustained in everyday life. An excessive orientation toward the present can even attract the suspicion of being merely an insistent, impulsive demand for short-term wish fulfillment. The hedonistic orientation toward stimuli is thus to be distinguished from the individual's personal ability to concentrate completely on the moment and not allow themselves to be distracted by outside stimuli.[7] It seems that it is indeed the individual who is not predominantly oriented toward the past (unable to let go), toward the present (impulsively reward-oriented), or toward the future (purely deadline-oriented), but rather who can switch freely between time orientations, who can exercise temporal freedom. We'll discuss this again later.

Of course, time orientations are interrelated. As the existentialist philosophers, for example, have shown, engagement with one's individual finitude and the acceptance of one's future death form the very starting point for being able to conceive one's life goals more consciously.[8] This consciousness of one's own destiny is the motivation not to just live unthinkingly for the moment, but to live such moments consciously. We are also the people we are *now* based on our past. I am conscious of myself as a person, nourished by the memories and stories I associate with myself. As someone experiencing in the present, I am characterized by my memory, by my life as a whole as a felt duration of time. As Peter Handke says in his poem, *To Duration*:

> duration is a matter of years,
> of decades, of our lifetime;
> duration, it is the feeling for life.[9]

We are our life as a whole.

The Duration of a Moment

Over the course of millennia in the history of philosophy, people have debated how to conceive of the present moment: as a mathematically describable point in the flow of time, as a simple intersection between the future and the past, or as an extended period of time.[10] The philosophical debate begun by Saint Augustine around the year 400 can be summarized as follows: if the present moment has a temporal extension then we can in turn determine a series of points in time on this notional temporal line. However, each of these points in time has other points in time that lie in front of and behind it. From the position of any desired point in time the other points in time represent what is past and what is future, as they lie temporally before or after.[11] In this case, however, this temporal line cannot represent the present moment, as it contains all three temporal dimensions. Accordingly, the moment must be without extension.

On the other hand, it can be argued that we still perceive movement and change, the experience of the temporal ordering of two events, as present. A shooting star burning up as a bright movement in the night sky, is experienced by us as a duration of time, albeit short. This lived moment of the shooting star is expanded; only because it is expanded can I perceive any movement at all. Experiencing music does not happen as a series of incoherent sounds. A melody is perceived as an extended unity. Even if the individual sounds in the melody have a temporal sequence, a Before and an After, they are integrated into an audible musical phrase, a whole. The musical experience depends on the perception of the extended gestalt of a melody.

This means, in physics and as an abstract philosophical idea, that the moment is a point in time without extension, but psychologically and phenomenologically it is an extended period of time in one's present experience.[12] Otherwise the perception of change and movement would not be possible. This is how we can interpret the proclamation of Goethe's Faust: "When to the moment, then, I say: 'Ah, stay a while! You are so lovely!'" "The moment" refers not to a point without extension in the continuum of time, but to an intensely lived moment in life that lasts. The lived moment is necessarily extended. Let us also remember that just saying the word "now" takes about a third of a second.

In these theoretical discussions, we must take account of two aspects of time consciousness:[13] the feeling of time passing and the experience of the present moment. On the one hand, we sense the passing of time. The anticipation of an event now becomes an event which ultimately and irrevocably becomes the past. This is the flow of time. On the other hand, I sense the presentness of experience. The totality of my consciousness feeds on the present-time experience of all my senses, my body awareness (my inner sense) as well as the external senses (visual, olfactory, and auditory impressions). In addition there are my thoughts, my memories as well as the plans I'm making for this evening. This is the internal and external experience of the present. This is what I am *now*: lived presence that must necessarily be extended in time, so that conscious experience of the temporally developing processes of the internal and external world is possible.

In order to better understand the lived moment and to be able to make it accessible to research, we can operationalize the moment, that is make it "quantifiable." In this we try to harmonize the subjective (first-person) perspective with the objective (third-person) perspective. This will become important for the further development of the ideas in this chapter. We are trying to record the phenomenon of the moment using the objective measurements of psychology and the neurosciences. This is a question of identifying and quantifying the processes that underlie the feeling of presence. One fundamental question is how long the felt, present moment persists—what is its duration. To do this, we can distinguish at least three different levels of the moment of experience, each of varying durations: (1) mental presence, (2) the experienced moment, (3) the functional moment. It will become apparent that the colloquial words that have so far been used synonymously, namely *moment, instant, now, present,* and *presence,* can probably be assigned to varying levels. The first two concepts, moment and instant, are better suited to the description of shorter durations of time, and the last two, present and presence, to longer durations of time.

An upper temporal limit applies to the duration of mental presence—the longest possible duration of our sense of being present. I can only describe as present what is now available to me as an internal or external experience. Anything that has just been experienced and thought and has not yet been forgotten is still present. In other words, all thoughts that can be operated mentally within the span of working memory are present. This span is essential for the narrative understanding of self, the stories I tell about myself when I activate my memories: who I am, how I have developed, and what I intend to do in the future.[14] Using experiments, psychologists can assess the span of working memory—the temporal window for the short-term retention of things such as numbers, words, and visual symbols. In this regard, independent of the situation or the nature and quantity of the stimulus, we talk about a short-term memory capacity of between several seconds and perhaps half a minute. But these are only rough guidelines. In the lived presence, feelings, thoughts, and sense impressions are integrated into a whole, forming me as active agent of a physical and mental self. The consciousness of the totality of the whole here and now—my experience of self physically and temporally extended—is *mental presence*.[15]

This account corresponds with the analyses of internal time consciousness developed by the phenomenologist Edmund Husserl (1859–1938).[16] All experience has an underlying temporal structure that consists of the components of *Urimpression, retention*, and *protention*. While the Urimpression is what has been perceived just *now* (let's say the now-heard sound in a melody), this current impression nevertheless remains simultaneously linked to the only just heard elements of perception, which slowly slide into the past as retentions but are still co-presently experienced as remembered components. In this way the now-heard sound is linked to the sounds heard earlier to form a melody. Protention, by contrast, is the anticipation of an event that is about to begin. If I am familiar with the piece of music, there is a powerful anticipation of what is about to be heard. The rules of musical harmony also co-determine what we think we're about to hear. We might define protention as openness to what is coming.

The fact that protention is not merely a mental phenomenon but is "embodied" in us is shown, for example, in the backward jolt we feel when stepping onto a stationary escalator. Our attitude of anticipation is geared toward the stairs moving upward.

Although we can clearly see that the escalator is stationary, the surprising effect of the felt backward movement occurs because we anticipated the opposite movement both bodily and motorically. In recent conceptualizations in neuroscience, perception has been described as a predictive, goal-oriented process directed toward the very near future. Perception as "predictive coding" means that based on prior experience the brain constantly makes predictions about what might happen. The goal thereby is to minimize surprise, defined as the discrepancy between prediction (what might happen) and actual sensory input (what then actually happens). This discrepancy essentially amounts to more or less "prediction error," which the organism aims to minimize in order to reduce energy consumption. If we are correct in our prediction, we don't have to adjust our behavior. Accordingly, normal waking consciousness can be associated with constant short-term predictions.[17] Intrinsically, the moment always consists of the three modes of time, which cannot be considered each on its own but form a mutually defining lived presence (see table 2.1). The musical note we are hearing now is influenced by what has already been played and by what is expected.[18] Through these three complementary components, time consciousness of the lived moment forms a temporal field.[19]

In the language of experimental psychologists, mental presence means that events that are initially anticipated, then perceived *now*, are eventually lost as memory traces, as the seconds are passing and new experiences are continually happening. Only a certain quantity of experiences can be kept present at any one time; very important experiences remain, but with time most disappear, as they can no longer be retained in the working memory.

Table 2.1

Three forms of present-time experience, their temporal duration, and the related phenomena and processes

Forms of present-time experience	Temporal duration	Phenomena, processes
Functional moment	Milliseconds: About 30 ms to 300 ms	Link between two or more events without perception of temporal order
Experienced moment	A few seconds: Between 300 ms and 3 or more seconds, approximately	Rhythmic units in the case of the metronome; frequency in the case of bistable images; synchronization of action and event
Mental presence	Between several seconds and a few minutes	Short-term memory Working memory Narrative self

Research also suggests another, shorter temporal moment. It seems to be the case that perception and action function optimally in units of up to around 3 seconds in duration (with some variation across phenomena). Perception is accumulated into units of this approximate duration, with the consequence that individual events are experienced as belonging to one moment in time. Acoustic temporal units, such as in the case of the metronome, where the individual beats are automatically assembled in groups of "one-two" or "one-two-three," exemplify this process of integration. If the intervals between two metronome beats are longer than approximately 2 seconds, the ability to integrate fails, and only individual, disconnected beats are heard. If the task is to accompany the metronome beats using one's own movements, the ability to accurately accompany the stimuli fails if the intervals between the beats are longer than 2 seconds.

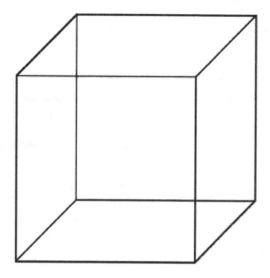

Figure 2.1
The Necker cube is a bistable image, which can be viewed "from above" and also "from below" and which can be used to investigate changes in the present-time experience.

In the case of visual bistable images such as the Necker cube (in which one can view a cube from two different perspectives; see figure 2.1) or the Rubin vase (in which one sees either a vase or two faces), the visual aspect alternates approximately every 2 or 3 seconds (*now* I see the vase, *now* I see the faces, etc).

As research carried out by the neuroscientist Ernst Pöppel has shown, a great deal of empirical data demonstrates how human perception and action are automatically grouped into units of this duration, and are thus associated with the feeling of present-time consciousness.[20] This temporal structure can also be found in language and music. Commonly repeated actions, such as people shaking hands in an everyday encounter or athletes and trainers hugging each other at sporting events, demonstrate cross-culturally an approximate duration of up to 3 seconds. Perception and action are integrated into a whole within a temporal window of this duration; they take place in these present-time moments. Both verbal and nonverbal interactions also take place in interpersonal, shared present-time units that have this temporal duration: shared experience is made possible by the synchronization of two or more people who are anchored in time and interact on this shared temporal platform of an *experienced moment*.[21]

In the 1950s, the artist Karl Otto Götz (who died in 2017 at the age of 103) developed a painting technique in which paintings could be created in an average of 3 to 4 seconds. Götz broke down his painting movements into three subunits of approximately 2, 1.5, and 1 seconds. Each of the three movements stands for a part of the creative process; all three are, however, connected in the larger unit of 3 to 4 seconds. Götz himself suggested that behavioral scientists and psychologists should research the processes that underlie painting. According to what we have already seen here, Götz's painting processes comprise the creation of present moments that form our experience and action. Karl Otto Götz devoted himself to his internal impulses toward movement in the "now" (see figure 2.2).

There also exists a lower temporal limit for the conscious perception of individual present-time moments, a limit that is defined by perception's capacity for temporal resolution. For example, metronome beats can only be grouped into rhythmic units if the individual beats are presented as not significantly faster than a frequency of 3 Hz; otherwise only a fast sequence of beats without emphasis is heard.

Figure 2.2
Karl Otto Götz, *Tunset*, 1958, Museum für Neue Kunst—Städtische
Museen Freiburg. Photo: Hans-Peter Vieser, inv. no. M87/012 © VG
Bild-Kunst, Bonn, 2015.

To the human ear, therefore, rhythmic units occur only in temporal intervals of between a third of a second and 3 seconds. Equally, it is no longer possible to synchronize finger movements with the metronome if the frequency exceeds 3 Hz and is therefore too fast. If people are instructed to press repeatedly on a key in their own personal tempo that is comfortable for them, what results is a "tapping" tempo of approximately 2 to 3 Hz. In the case of a maximum "tapping" tempo, carried out by people with a movement frequency of between 5 and 7 Hz, one does not experience a temporally individual key-pressing, as these movements happen too quickly.

This lower limit for conscious present-time experience in a series of events indicates that in the case of shorter intervals the temporal sequence can no longer be perceived. As experimental research shows, there are certain interval limits below which the temporal order of events cannot be detected. For two short stimuli, temporal order thresholds in the region of 20 to 60 milliseconds seem to exist across modalities. If two auditory or visual events take place at an interval below this threshold, a test subject cannot recognize the temporal sequence—the two events are blurred and cannot be placed in a temporal order.[22] In order to be able to detect a sequence of three or more acoustic or visual stimuli in their correct sequence, the individual events must take place at intervals of at least 300 milliseconds.[23] These thresholds provide clear proof of the temporal windows of perception, which integrate events in the environment into *functional moments*. Within these windows no temporal order can be detected. The stimuli are combined into present-time moments without the perception of a temporal sequence.

If we talk about an experienced or a lived moment, then we mean the moment of just a few seconds that structures our perception. This is the moment of the "present-time" experi-ence. This experienced moment is embedded in the mental presence of my self, as a narrating, commentating self. It is the consciousness of my self as a perceiving self.

Mindfulness and Time

What impact does heightened attention have on the moments of experience? Does subjective time actually expand through intensified present experience? In the aforementioned sense, a higher degree of mindfulness in experience would on the one hand influence the felt moment, and on the other hand lead ret-rospectively to time expansion. When I experience more inten-sively now, I also have more memory content. From research in experimental psychology we know that a greater concentra-tion of memory content stretches experienced periods of time. By contrast, temporal intervals that are less filled with memory, because of the monotony of always experiencing the same things, appear much shorter—simply not lived.[24]

Being "mindful" in everyday language means taking care in what is happening, a particular kind of attention—the opposite of carelessness or inattention. The concept has gained a fur-ther meaning in the last few decades through the emergence of notions associated with "mindfulness." The medical pro-fessor and teacher of mindfulness meditation Jon Kabat-Zinn describes the altered experience of time through the attainment of mindfulness in daily life as follows:[25] "The ... way to slow down the felt sense of time passing is to make more of your ordi-nary moments notable and noteworthy by taking note of them." In relation to the passing of time in the moment: "The tiniest moments can become veritable milestones." And in retrospect:

"Your experience of time would slow time down." The notion of mindfulness, as Kabat-Zinn uses it, has its origin in Theravada Buddhism. In Pali, the Central Indian literary language, the corresponding concept meaning "mindfulness" is *sati*, which means "awareness of the moment." The word is derived from the verb *sarati*, which means "to remember."[26] As we can see, this derivation of mindfulness, on the one hand from the experienced moment and on the other from memory, corresponds with our temporal concepts of lived, present time and retrospective, remembered time. If I am more intensely aware of the moment, then I also can remember better. This is how time expands.

Mindfulness can be exercised using forms of meditative practice. In the context of Buddhist traditions, meditation is embedded in wide-ranging spiritual teachings and practices. Due to cultural transfer and secularized usage, mindfulness meditation in the West concentrates on the method of experiencing the present moment consciously without judging or making value judgments.[27] Accordingly, the two central elements of mindfulness practice are (a) presence (the awareness of the present moment) and (b) acceptance (refraining from judgment). These two aspects are inextricably linked in the practice of mindfulness exercises. While the meditator opens up to the present moment and concentrates, for example, on his or her breathing, thoughts and feelings come and go. If one were to attend to them, one would lose the experience of presence. The world of thought keeps us separate from the present experience. Ultimately, through these exercises, one learns how to deal with one's emotions, as one gets to know them better and no longer pushes them away. One is no longer helplessly at the mercy of emotions when overcome by anger or embarrassment, for example, because the accepting distance from one's emotions helps to disrupt the automatic emotional reaction.

The acceptance of one's own thoughts and emotions also underpins the ability to focus on the present moment; with time, control over mind wandering increases. The process as a whole can be called the self-regulation of attention. This ability lies in not being distracted by emotionally freighted thoughts—"that was a great evening with friends yesterday"; "tomorrow I must get round to doing that darned tax return"—but rather returning to the fixed anchor of the present moment in the event of mind wandering.[28]

In this context, the choice of the anchor in the present is in no way random. Directing attention to the body and physical processes (such as breathing) supports the novice as they maintain their orientation toward the present. Through the sensation of one's immediate physicality, the feeling of presence is heightened. Ultimately, we are always anchored in the here and now by our embodiment. In focusing on the body, the conscious self is inseparably linked to the feeling of present temporality as embodiment extended over time.[29] Over the course of time and with increasing practice, if the act of focusing takes place more and more effortlessly, other kinds of attention regulation can also be chosen, culminating in free-floating attention paid to perceived moments without focusing on an explicit object.[30]

Mindfulness is an individual capacity, but the ability to practice it can be acquired through meditation techniques. For the science of consciousness, both aspects are of great interest, as they allow us to understand the conditions for measuring the experience of time and the present moment. As research into mindfulness meditation shows, various components of attention improve through the regular and persistent practice of meditation. While many effects can be attained only though continuous practice over a period of years, some of the positive changes appear even after just a few days, as scientific studies show: the capacity (a) for sustained attention, that is, to react quickly and precisely to different visual stimuli over an extended period of time of approximately half an hour,[31] and (b) for focused attention, that is, not being distracted by disruptive stimuli.[32] Standardized computer tests showed that attention regulation—a central element in the mindfulness method—is improved after intensive meditation training. In this case, the working memory span is of particular interest, since it provides a metric for mental presence. Standardized tests are used to assess the ability to operate using abstract symbols for a duration of more than a few seconds—that is, not immediately forgetting them. Studies using such tests have shown that working memory performance can be improved through mindfulness meditation.[33] On the whole, the research results consistently indicate that as a result of meditation training focused attention ability is heightened, mind wandering during assigned tasks is reduced, and working memory capacity is increased.[34] To this end, mental presence is increased through meditation techniques.

Systematic research into mindfulness meditation and time perception is still in its infancy. Just twenty years ago research into consciousness was generally frowned upon. Systematic scientific research into the effect of meditation on consciousness has existed for just over ten years; since then studies have been published even in mainstream scientific journals. Regarding effects on the experienced moment, an important first step was made by a group of researchers whose results were published in 2012. The question was formulated in the title of the study:[35] "How Long is Now for Mindfulness Meditators?" Thirty-eight experienced meditators from various schools, all working with a focus on mindfulness, were compared with 38 people who had no experience of meditation techniques. The minimum experience required was regular meditation over a period of five years. Under relaxed conditions, while looking at a Necker cube that can in principle be seen from two aspects, each person was asked to press a button every time the perspective changed.

On average, a change in perspective happened every 4 to 5 seconds. In this respect the two groups did not differ significantly from each other. In a second round, the test subjects were asked to "hold" the perspective for as long as possible before it changed "of its own accord." In this case, a clear and statistically significant group effect occurred. The meditation group could hold the perspective of the Necker cube for almost 8 seconds on average, while the control group could only hold it for 6.2 seconds. This research was an initial study, which clearly leaves many unanswered questions that must be explored further. If the length of time of seeing one perspective of a bistable image is linked to the experienced moment—as presumed in theory— then these results suggest that experienced meditators can expand their experience of the moment at will.[36]

Further proof of the thesis that focusing on the moment leads to subjective time expansion was provided by a research group at the University of Kent in England. In this study, a group of people without any specific experience of meditation was played a 10-minute CD recording of a meditation induction through concentrating on the breath, while a control group listened to a 10-minute extract from an audio recording of *The Hobbit*.[37] In each case, before and after the relevant intervention (meditation versus audio recording), the test subjects estimated durations of time in the range between 400 milliseconds and 1.6 seconds. A relative subjective time expansion effect was shown in the meditation group but not in the *Hobbit* group. This is a very surprising effect. The test subjects were students without evident experience of meditation, and a 10-minute meditation was enough to achieve a relative subjective lengthening of duration. Evidently, for a short time after the meditation the students entered into a more mindful state, leading to subjective time expansion.

In a study of 63 students carried out in Munich and Freiburg from 2012 to 2013, the link between mindfulness and the experience of time was surveyed.[38] Questionnaires—tools for self-assessing everyday mindfulness—were used to measure individual mindfulness in everyday life.[39] The questions concerned the awareness of external events, of consciousness of the body, and of emotions. As discussed before, mindfulness addresses conscious presence in the here and now with an attentive, accepting attitude toward events, that is without immediately judging events as "good" or "bad." One of the two questionnaires used in the study was the Freiburg Mindfulness Inventory (FMI) created and validated by Harald Walach, Stefan Schmidt, and their colleagues.[40] The questions in the FMI can be organized according to the two main components of mindfulness, "presence" and "acceptance." Those with a high degree of "presence" are "open to the experience of the moment," "perceive how emotions are expressed in their body," or "feel in their body when eating, cooking, cleaning or talking"—that is they have a lived, physically conscious experience of the moment. A high degree of "acceptance" is demonstrated by those who are "in contact with unpleasant, painful sensations and emotions" or who are "conscious of the fleeting nature and transience of their experiences." As a second questionnaire on the understanding of mindfulness the Comprehensive Inventory of Mindfulness

Experience (CHIME), developed by Claudia Bergomi, Wolfgang Tschacher, and Zeno Kupper, researchers at the University of Berne, was used.[41] Using simple, everyday questions, the inventory identifies eight dimensions of mindfulness, which are related to internal and external experiences, to the openness to experience, and to acceptance and being nonjudgmental.

The results of the study indicate that students who consider themselves more mindful are more accurate and emotionally stable in the temporal organization of their behavior. The students who describe themselves as more mindful are less impulsive and more concerned with the future.[42] These results are only contradictory at first glance. Shouldn't more mindful people be more focused on the present moment and thus less oriented toward the future? These results can, however, be interpreted in terms of greater temporal freedom. Those people who consider themselves more mindful are more conscious of their perception in the moment and of their emotions. Accordingly, they are more strongly oriented toward the present. This is the ability to experience the moment intensively. But they do not "cling" impulsively to the now, expressed as a drive for immediate gratification. More mindful people are less oriented toward stimuli, that is, they can alternate between controlled attention to present experiences and necessary planning for important future events, as required.[43] This is precisely the difference between stimuli-oriented attachment to presence with a lack of freedom, and the ability to focus and give stable attention to present experience.

Interestingly, in time estimation tests the students who describe themselves as more mindful are also more accurate when judging duration in the temporal range of milliseconds and several seconds than those who identify as less mindful.

They identify smaller differences in temporal duration between two acoustic stimuli and more accurately reproduce temporal intervals by pressing a button, doing so with smaller deviations from the specified duration. This ability can be attributed to stronger control of attention given to the external stimuli to be estimated. Mindfulness training improves performance not just in time perception tasks, but also generally in many perceptual and intellectual tasks.[44]

These research results account for the changes in the experienced present moment that occurs in people with heightened mindfulness. From these initial results we can infer that heightened mindfulness and presence leads to an expansion of subjective time. Subjective time is modulated by functional states of mindfulness. These states can be described as (1) attention regulation (remaining in the here and now; no mind wandering), (2) consciousness of the body (awareness of bodily self), and (3) emotion regulation (nonavoidance and acceptance of emotions).[45] These three functional states are also causally linked to the modulation of temporal experience:[46] (1) A heightened attention to time leads to an expansion of subjective duration. We often become painfully aware of this during periods of waiting, when we are watching the time, which passes so very slowly toward the longed-for arrival of the event. (2) Body awareness: When searching for the mechanisms of time perception, physical processes were identified as a possible factor in the subjective feeling of time. For example, under sensory deprivation in the floating tank, where one floats in body-temperature salt water, one neither sees nor hears anything. Even under such circumstances the immediate consciousness of the passing of time remains intact. Indeed one's own physicality cannot be turned off, and the experience of one's own time becomes more intensive. (3) Through the admission of emotions the sense of presence of the felt moment increases, resulting in a heightened consciousness (of self).

If a heightened presence results in moments that are experienced more intensively, then time should expand retrospectively as well. If I look back over my life as a whole, the experiences create a feeling of duration. According to the observations discussed here, more mindful people should experience a lengthening of previous periods of time. Accordingly, life as a whole would expand. The first empirical proof of this to the extent of minutes was provided by Romanian psychologists in 2013.[47] A group of students underwent 30 minutes of mindfulness training every day for a week. In a test situation, this group estimated the subjective duration of two entertaining BBC documentary films lasting exactly five minutes, and the waiting time preceding them. Only after the film ended were the test subjects asked about their felt time. The group of students with mindfulness training estimated both periods of time as passing subjectively slower than the control group of students who had not undergone the training. Thus the period of time expanded retrospectively for the meditators by contrast with the nonmeditators.

But what happens in the case of really long periods of time that I experience as my lifetime? In this case too, a study completed in 2014 and conducted in Freiburg and Munich by Stefan Schmidt, Karin Meissner, and me showed clear results.[48] Involved in the study were 40 women and men with many years of experience in mindfulness meditation. On average they had been meditating for 10 years, and in the last eight weeks they had regularly meditated for seven hours every week. The selected control subjects without experience of meditation were of the same gender and the same age (on average 40 years old) and had a comparable education. Through simple questions, they were asked about their sense of the course of time in general (how time passes), and about the previous periods of a week, a month, a year, and the last 10 years. In addition, several questions also elicited information on the everyday experience of time pressure and the feeling of time expansion.

The results can be summarized as follows: experienced meditators experience less time pressure and feel more time expansion. These findings correspond with the general sense of time passing more slowly for the meditators. When asked about temporal intervals in the past, they gave evidence of time expansion in the relevant periods of their lives—in particular during the last week and the last month—by comparison with the nonmeditating people we surveyed. According to these results, life as a whole passes more slowly and past periods of time expand for people who live mindfully.

Beyond the Moment to Timelessness

People who have experienced extreme manifestations of extraordinary states of consciousness often report a radical alteration in their sense of time. As we saw in chapter 1, this can occur under the influence of hallucinogens—mescaline, LSD, or psilocybin. People with near-death experiences recount afterward that time and space had lost all meaning. In rare moments of mystical experience this event can overcome an individual—the feeling of being at one with the world while at the same time space and time dissolve. These experiences are very important for understanding the formation of consciousness of self, for they show how closely related our consciousness is to the perception of time, space, and body. The science of consciousness can learn a lot through controlled studies of these aspects of extraordinary states of consciousness.

To this end we can also interview and test experienced meditators. As a result of years spent practicing contemplation, their self-perception is particularly heightened. This ability makes it possible for scientists to combine their introspection with behavioral data metrics and the related brain activity—in the truest sense of the word, neurophenomenology.[49] In other situations, introspection is considered an unreliable method because test subjects often have difficulty verbalizing their experiences. By contrast, in this context people with decades of experience in meditation techniques, and, consequently, heightened attention and self-perception, can provide valuable insights.

In one study, the Israeli researchers Aviva Berkovich-Ohana and colleagues looked at twelve experienced mindfulness meditators (with an average of 16.5 years' experience), examining them while they were in a meditative state and measuring their brain activity using MEG (magnetoencephalography).[50] Brain activity in particular within the theta band (with a frequency of between 4 and 7 Hz) was affected by various states of consciousness. During meditative states, activity in areas of the brain that are linked to the processing of the sense of body increased. The respective states of consciousness were described phenomenologically in retrospect by the meditators as, for example: "the sense of time and space became hazy," "time lost its linear form," or "time didn't exist." As these and other descriptions show, the sense of time and timelessness varied among the meditators. Perhaps only a few will have had a "real" feeling of timelessness (see our discussion below). But this study is a further building block in the research and demonstrates how activity in areas of the brain that play a role in the sense of body is modulated by the conscious alteration of time.

Meditation instructors teach that progress toward the sense of timelessness is made by concentrating on the present moment. An aptitude for contemplative meditation that has been obtained through training over many years leads to states that are hardly ever accessible to people without such training. In everyday life we experience our body situated in space and moving through time. In this context, are space and time, the two basic experiences of "pure intuition"—as Immanuel Kant called them in his *Critique of Pure Reason*—supposed to simply dissolve? We asked the meditation teacher Tilmann Lhündrup Borghardt, who can look back on 35 years' experience of spiritual meditation.[51]

Together with Stefan Schmidt, who is engaged in the scientific study of mindfulness meditation at the University of Freiburg, I met him to discuss this. For 21 years Lhündrup Borghardt has lived the life of a Buddhist monk in a monastery in France, during which time he has meditated for 12 hours every day for 10 years.[52] During the other years as well, Lhündrup Borghardt meditated for several hours every day. Thus, over the years, he has accrued a total of approximately 50,000 hours' experience of meditation. Today he teaches meditation in several countries. In our conversation we talked about understanding the altered experiences that a master of meditation such as he can consciously induce, in relation to the sense of time and the experience of self. In particular our topic was timelessness and the dissolution of self. Tilmann Lhündrup Borghardt's ideas (TLB) are reproduced below in italics. The state of "awakening" describes the highest stage in the spiritual experience of meditation that one can attain. In these discussions we touched upon many aspects, which cannot all be described here exhaustively. The focus is on the self and its time.

TLB: *Timeless awareness during meditation is an awakening. It has neither beginning nor end. This timeless time is plunging into a being in which no comparison takes place. In comparing there are always relationships between before and after. It is timeless presence without a sense of self, without observers. Perception and perceiver are one. It is about merging into the visual or auditory impression. You lose yourself in hearing and seeing, as the experience of hearing and seeing needs no observer, no self.*

TLB: *But how do we know that this state has been attained? If this experience is described as a process, then that is not the state of absolute presence, as time exists in it. For example, there is the mystical experience, being at one with the world, that approaches the state of awakening. It is a liberating feeling of spiritual peace. But an observer is present (a self), who perceives the migration of the sun or the movements of animals (time). This peaceful and relaxed experience still contains the dual consciousness (subject-object split). Ultimately a "self" experiences that they are at one with the world. Thus this sense of unity still has an observer. This kind of mystical feeling can also be produced by drugs. In the preliminary stages of meditation, too, one is astonished by the mutability of time, which can however still be experienced.*

According to this, the mystical experience with a subject (self) is a preliminary stage of timeless awakening. In the mystical experience the observer is still located in space and time. However, he or she has the feeling of merging, of being at one with the world. By contrast, in the state of being completely at one with the world, the self and time disappear completely. There is only perception, not the perceiving individual.

TLB: *In awakening observers forget themselves, as they merge into perception and the duality of subject and object disappears. The knowledge of selflessness and timelessness only happens in retrospect; it is accessible subsequently. The timeless state cannot be described better than through the concept of "timelessness." Being at one with space is equally difficult to describe; it is being without a middle and borders. If we integrate this experience into everyday consciousness, a "vibrating basic awareness" is perceived, a feeling of living presence arises, a preparedness for sensual experience without entering the empirical world. The awareness lies in itself. There is no seeing, although our eyes are open. It is as if perception is transparent, all is the space of awareness, all is pure consciousness.*

This description recalls Immanuel Kant's epistemological concept of "pure intuition." The pure intuition of time and space exists before any experience (empirical world), and it structures sensual experience both temporally and spatially (see figure 2.3). To a certain extent pure sense data are embedded in a temporal system and in the three dimensions of space.

Essentially our experience contains both components, the empirical and pure intuition, content and form. As TLB describes it, the pure spatial experience in the state of contemplative "awakening" is possible without reference to the empirical contents of sensory experience. Yet this pure experience is neither temporal nor spatial.

TLB: *The orientation of seeing and hearing is there, without a substantive visual or auditory impression being present. In addition the feeling of self is missing; it's an experience without a center. It is immediate perception without emotional and cognitive filters. Normally, there's an experience of self that absorbs a lot of energy, the wishes, ideas, and hopes that we cherish (the filters of perception). In the state of awakening one is in harmony with the situation without*

Figure 2.3
This picture of a sunset over the sea recalls paintings by Mark Rothko. In his paintings, Rothko tried to visualize meditative experiences in which time and space dissolved.

being related to the center (without self). One can imagine this in stages, being a bit like the experience of flow, when one merges completely immersed in one's activity.

In the experience of flow, when one is completely given over to an activity, whether writing, playing music, or sport, the experience of self is greatly reduced. In the flow one also forgets the time and is surprised afterward how much time has passed.

TLB: *What we learn from awakening it that there is no self-center in perception and action. Anyone who has ever had this experience can no longer go back to the old belief in a self as the center. In this context we should perhaps make a distinction between the individual "center-self" and the "networked self." In the notion of the center-self we proceed from an essence of being which is "self." This "self" wants, hopes, and desires. This is the personal self, which makes emotional and cognitive assumptions about the world and interprets it. In this, people are very different. The networked self, in which people are very similar, is not a self in the individual sense. This is about the self-functions of responsiveness and empathetic behavior, of the capacity for love. These capacities are in us all; we are not so very different from one another in this regard. There are parallel qualities of a net-worked self without a self-center which allow us to do the right thing in harmony with the world. These qualities can have an increasingly free and strong effect, when not so many blockages exist anymore. The fixing sense of self dissolves in the flowing functioning of the interplay of innate qualities. The individual then becomes more free from anxi-ety, more concentrated, more related to reality, and more empathetic. Very importantly, it is not a self that exists in reality but the idea of a concrete self that dissolves.*

TLB's accounts with regard to notions of self complement ideas put forward by the philosopher Thomas Metzinger, who proposes the thesis that there is no center-self, as his book title explains: *Being No One: The Self-Model Theory of Subjectivity.*[53] Metzinger's view is that there is no static "self" as a permanent entity or substance. Rather, he says, mental processes continu-ally create a self-model, the illusion of a "center-self." Driven by bodily signals, this is a physical and spatial anchor for the phenomenal self. Brain research shows that alongside a sequen-tial processing of bodily and environmental stimuli there exist many parallel lines of processing without any localizable inte-gration center. Rather, the parallel, spatially distributed neuro-nal processing modules are linked together to form a whole, in which they are coded as operating "simultaneously." This present-time linking of the processes might be the mechanism that ensures the conscious experience of a self in the present moment. Through this we have an idea of a self, a self-model of the "I."

TLB: *The way to awakening is via presence in the moment. It is the open awareness without controlling entities that evaluate and judge and thereby create distance once again. My experience with meditation groups is that people find access to this open awareness relatively quickly. It is an atmosphere, which enables you to experience trust and to loosen the inhibiting control mechanisms. When that happens, the way to presence is via the six senses. But in presence something else takes place. It is a change from "what happens" to "how it happens." The "what" orientation refers to the object. The "how" orientation refers to the quality. How does what I experience have an effect on me? A question I often put to students is: "What is being like?" Not, "how is it for you?" but "how is it now?" Through the presence of the "how" orientation without an observer the presumed self is suffused with awareness. The perceiving individual and what is perceived become one. Being is then clear and present. It is nothing that I might grasp. It is inconceivable. Being. It is the emptiness of which Buddhists often speak. This constitutes the timeless quality. In every experience there lies the timeless dimension, the quality of awareness—even now as we speak and hear. That is the mystery of life. But it can't be conceived in words. When the spirit dissolves in this inconceivable quality—that is awakening.*

In our conversation, Stefan Schmidt brought up the following idea: experience is primarily timeless. Only the introduction of an observer introduces time. This is the ability to distance oneself from the now, to be mentally elsewhere, to compare. This leads to the creation of time consciousness. TLB added: *We need the sense of time, because we want to communicate, because we make plans; it is something meaningful, but it is not already part of experience.*

Certain species of animal also have a time consciousness. These are the species to which we also attribute self-consciousness and a theory of mind. Theory of mind is the ability to adopt the subjective perspective of a fellow member of your species. Crows, great apes, elephants, dolphins, and whales have a consciousness of themselves and they also plan for the future.[54] They are able to disengage themselves mentally from the present situation and wait for an event, if it brings them advantages. The self-conscious observer compares the present situation with one imagined in the future. Research proves the close link between consciousness of self and time consciousness. Phenomenologically oriented philosophers also argue that the self is created in reference to the future and the past. I am conscious of my self through my memories of my self and through my plans.[55]

When we begin meditating, we become particularly conscious of time. We have no distractions; we are completely concentrated on ourselves and the present situation. Bodily presence becomes particularly clear to us when we adopt the posture of meditation. We are completely body and self. In this situation, time passes quite slowly for the beginner—also because we perceive ourselves in reference to the future, knowing that we must remain in this position for 15 more minutes.

TLB: *The more one gives up the need for control and orientation, the more intensively one enters a region where time no longer plays a role. Time plays an important role for people, as it is concerned with control. If I no longer pay attention to time, then I no longer pay attention to the self in time. If I am completely relaxed, if I really don't care how long the meditation lasts, this is the best requirement for entering deep contemplation.*

Edmund Husserl's phenomenological analyses employ the term *Urimpression*, or primordial impression, introduced early in this chapter. The *Urimpression* of the present-time experience is always embedded in retention (awareness or immediate memory of what has just happened) and in protention (the expectation of what might be about to happen). In this notion, we cannot understand the *Urimpression* separately from retention and protention. What we experience "now" is always already dynamically interwoven with the two other modes of time. The sound that resounds now is influenced by previously heard sounds, and these units of sound influence my anticipation of sounds to come. The perception *now* is characterized by this expanded temporal field. According to phenomenology, I am not in a position to disengage myself from this temporal field; the flow of time of what is about to happen and what has just happened are part of my present perception.[56]

Against this backdrop, meditative awakening can be described as immediate contact with the *Urimpression*, in which the flow of time doesn't occur. It is the direct contact with the *now* of perception. Perception is completely in the now and without an observer, without a self.

A dissolution of body boundaries during meditation actually can lead to greater happiness, as a recent study by the French psychologist Michael Dambrun has shown.[57] The study participants were not experienced meditators but regular students at the university. In the study, they either followed a 21-minute audio tape with a body-scan meditation instruction or just spent the same time resting. In comparing the two groups it became clear that those students who had meditated felt their body boundaries to a lesser degree and they felt happier afterward than the control group. During a body-scan meditation one is guided to focus successively on different parts of the body. Initially, one feels a stronger sense of the body—and of the passage of time. But after a while one loses the sense of bodily self and of time, entering a meditative flow where the feeling of a self is less dominant and happiness increases.

How can one understand the link between less selflessness and happiness? In our pursuit of happiness, we are often self-centered. I want this, I want that. This hedonistic principle can lead to pleasurable states when I am able to consume. But it can also lead to unpleasant situations when I do not get what I want. The striving for personal pleasure therefore leads to fluctuating states of happiness depending on the contingencies of life—factors I may not be able to control. Relying too strongly on these self-related but not necessarily controllable external rewards can easily lead to unhappiness. In their recent happiness model, which they call the Self-centeredness/Selflessness Happiness Model,[58] Michael Dambrun and Matthieu Ricard argue that self-centeredness develops when we take our own condition to be more important than the condition of others; the self is experienced with sharp boundaries and as separate from the others and the world. A more selfless existence, in contrast, is based on the feeling of a weaker separation of oneself from the surrounding world and a greater connection with other people. Selflessness also comes with beneficial emotions such as compassion and love. The self-centered individual is a static self with rigid desires and regularly occurring disappointments. The "selfless self" is engaged with others and results from a dynamic state of acceptance of what comes and goes in life—resulting in more happiness.

Daydreams and Mind Wandering

One major difficulty for beginners and advanced practitioners of meditation are thoughts that appear and persist, disappear again and are replaced by new ones. Thoughts about yesterday and tomorrow interrupt our attention on the here and now, on spatial and temporal presence, as does losing oneself in potential worlds and fantasies. Just as we were completely focused on our breath, the very next moment thoughts about tomorrow's meeting can start to unsettle us. Then other thoughts intrude, and suddenly we're lost in entire trains of thought. Only many seconds later do we become aware that we have wandered off, and we return to concentrating on our breath.

By comparison, the art of meditation is about maintaining presence. For example, some kinds of meditation use a mantra, a syllable, or a word that is repeated out loud incessantly. This functions in a similar way to praying with a rosary: by reciting the Our Father, the Hail Mary, and the Glory Be to the Father in the sequence of the prayer beads, one heightens the capacity for absorption in meditative prayer, ensuring that disruptive thoughts do not even reach consciousness.

In everyday life, too, mind wandering at the wrong moment can be disruptive. Those who cannot concentrate on an important task are slower to accomplish it and sometimes don't reach their goal. The ability to direct one's attention to the point is just as essential for housework as it is for repairing cars, for example, and for any kind of observation, whether its object is a work of art in a museum or a sunset. Heightened mindfulness in life, with the accompanying patience, has a quite mundane utility, whether it's a question of perception or of problem solving.

This does not in any way mean that daydreams and mind wandering are not essential for our lives. On the contrary, they are a cognitive mechanism that leads to creative ideas, namely relinquishing focused attention. Many of the ground-breaking ideas of scientists, artists, engineers, and product developers appeared in moments of mind wandering and daydreaming, that is precisely when they were dealing with the particular problem while they were not concentrating.[59] Experimental studies by the American psychologist Jonathan W. Schooler and his colleagues revealed[60] that the contents of daydreams are indeed predominantly linked to goal orientation and future planning. Furthermore, if, having been set a very difficult task, test subjects were subjected to a monotonous activity in between, they were subsequently able to go back and solve the difficult task more often than if they had been asked to pursue another focused activity in between. That means that during the monotonous activity, the right ideas for subsequent problem solving were developing "beneath the surface." Unfocused mind wandering is a mode that sorts through one's thoughts and produces solutions. If we can't find the solution to a problem, a stroll often helps. A walk around the block or in the woods is at the same time a way of giving our thoughts a walk. Often the answer comes to us suddenly, like divine inspiration (as was thought in earlier times) or as a manifestation of unconscious processing (as we interpret it today). As is clear from these examples, mindfulness and mind wandering or daydreams do not play off against each other. It is a question of balancing both factors, in each particular situation: being mindfully focused here and now; and letting the imagination take its course in daydreams and surprising ourselves with ideas.

3 Loss of Time and Self

A Puzzling Patient

In 1905, the French psychiatrist Gabriel Revault d'Allonnes published a scientific essay on the subject of a patient at Sainte-Anne psychiatric hospital in Paris: *Rôle des sensations internes dans les émotions et dans la perception de la durée* [The role of internal sensory perceptions in the emotions and perception of duration].[1] In the essay's very title Revault d'Allonnes interweaves "internal sensory perceptions" with the "emotions" and the "perception of duration." In the case of the patient, Alexandrine, these three mental functions have been impaired. The 53-year-old hospital patient—she is married and has a son—complains that for about a year she has had hardly any physical sensations, experiences no emotions, and no longer senses the passing of time. Revault d'Allonnes records his conversations with the patient Alexandrine in great detail in order to explore this strange case in depth. Although she cries frequently—tears well up if she has to think about a personal misfortune, and even during her conversation with the doctor—in doing so Alexandrine experiences no emotions. This means that the physiological reactions to and the cognitive understanding of a situation appear intact, but the subjective side of the affective experience is disturbed:

—Alexandrine: You see, Sir, I can cry! But it doesn't affect me, I feel nothing. ...

—Revault: But you are sad.

—Doubtless, as I am indeed crying. I have reasons to be sad, my illness, being separated from my poor husband, and from my son, he is in such poor health! No, I do not lack reasons to be sad. I am crying but it doesn't affect me, I can no longer feel anything.

Physical sensations are also affected. Alexandrine has certain physical perceptions, but is not particularly affected, for example in the case of feeling cold:

—You are shivering.

—Perhaps my feet are cold, but that doesn't bother me; I feel the cold a little, but it doesn't affect me. In the sun I feel the warmth a little, but it doesn't affect me.

Equally, Alexandrine is never hungry, and never senses when she is full. She reports that she has to manage her food intake using her experience, since after all she knows approximately how much a person eats. She is seldom thirsty, is never tired, and also does not feel rested. The urge to visit the toilet is absent. Revault d'Allonnes finds that her descriptions fully correspond to reality through a series of tests, in which for example he pours ice-cold water on sensitive parts of Alexandrine's body, provoking no increased sensory perceptions in the patient.

In his essay, Revault d'Allonnes discusses the possible causes of the disorder. He wonders whether perhaps hysteria, a psychiatric diagnosis common at the time, would be applicable. In hysteria symptoms, a lack of physical sensation could also present. Tests in which the skin on Alexandrine's body is systematically pricked by a needle show an extensive absence of skin sensitivity. Revault d'Allonnes contends, albeit rather weakly, that the illness must have an organic cause, provoked multicausally by several factors. In particular, he cites a serious case of influenza and complications during the menopause. Today, from a psychiatric point of view, we would probably talk about a case of depersonalization.[2] Incorporated in this diagnosis, he says, are two episodes of depression lasting several months, which occurred well before the beginning of the present disorder. However, one would, he says, have to determine neurological causes as well. For example, the patient reports serious headaches, possibly symptoms of a brain illness that directly preceded this puzzling condition.

Independent of the medical causes of the illness, there are symptoms that may be somewhat surprising, but which are crucial for Alexandrine, and which are questioned intensively by Revault d'Allonnes: problems with the perception of time. Since the start of the illness a year earlier, Alexandrine can no longer sense the passing of time. This is the immediate feeling of time. She can, of course, name the date correctly by looking at the daily newspaper, and she can tell the time on a clock, so she does not arrive late for her hospital appointments. She understands the abstract concept of time. But without the help of the clock she is far from being able to judge temporal duration. If we have to estimate a period of hours without using the clock, almost everyone will get it wrong by many minutes. But we have an approximate sense of the time that has elapsed, even if we make a clear error of estimation. However, this patient has absolutely no sense of how much time might have passed. Interestingly, it is Alexandrine herself who makes the connection between the loss of her subjective time and the disappearance of her bodily feelings. For example, she realizes that in the past she was able to judge the time by her rising feeling of hunger, her urge to urinate, or her tiredness. She herself infers from the situation that, because she no longer has any physical sensations, as a result she can also no longer sense time. Revault d'Allonnes summarizes his investigations thus, that "the duration perceived by consciousness is nothing less than visceral sensibility." He writes that we have something resembling internal clocks made up of various physical rhythms, supplied by signals from our gut, bladder, lungs, arteries, and heart. Consequently, physical perception is the necessary and sufficient condition for the immediate temporal consciousness of "medium-sized duration," as Revault d'Allonnes puts it. Longer periods of time he considers to be a day, a week, or a year; that is, intervals that cannot be experienced immediately by means of their duration. Medium-sized durations of time, by contrast, are seconds and minutes, up to perhaps a few hours—periods of time that are accompanied by physical reactions.

With the aid of a metronome, Revault d'Allonnes also conducts tests on the experience of rhythm. Quite short durations, of one second at most (at a frequency of 1 Hz), are employed. In several sequences, he sets the metronome beats at various frequencies between 1 and 3 Hz. After performing a specific sequence of 10 metronome beats at speed A, he presents, for example, a second sequence of 10 metronome beats with a different speed B. Alexandrine has to say whether she can hear the difference in temporal intervals. It is a proper experimental setup. And by comparison with healthy control subjects, the patient turns out to be impaired in the case of these short temporal intervals; she needs greater differences in pace in order to recognize them. But, Revault d'Allonnes explains, the impairment is not as serious as in the case of medium-sized durations of time, where a complete failure in the sense of time occurs. In a further test, Alexandrine is asked to match the rhythm of her breathing to the slower metronome beats; she fails in this simple task as well.

Body, Emotions, Time

At the end of the nineteenth century, William James and Carl Lange were already arguing that emotions were based on physical reactions. The extreme interpretation of this posited connection—"I perceive certain physical reactions and only because of this do I have a certain emotion"—is incorrect, but all our emotions are underpinned by physical movements (arousal versus relaxation), which feed into the context of the situation and participate in the expression and strength of the emotion. Externally, this is apparent in an altered physical posture, a high flush to the face, a strained voice, breaking out in a sweat, and flashing eyes. We ourselves experience our accelerated heartbeat, an uneasiness in the pit of the stomach, and butterflies in the belly—or, as Chuck Berry sang, "I got lumps in my throat when I saw her comin' down the aisle, I got the wiggles in my knees when she looked at me and sweetly smiled."[3] Emotions are physical. But these bodily reactions do not have to be conscious. Our emotions are quite directly what they are—joy or anger, desire or torment. For example, we do not have to feel physical sensations first, from which we then derive emotions.[4] The sympathetic (arousing, activating) and parasympathetic (relaxing, inhibiting) autonomous nervous system regulates the body's organs and its functions—the vital functions, heartbeat and breathing, blood pressure, metabolism and digestion, and of course, sexuality. The physical processes are to a certain extent integrated into complex situation-dependent emotions; as physiological reactions they are part of the emotions. The subjective experience of emotion can include the bodily experience, but not necessarily so. Physical processes, however, are always part of the emotions.

Today we talk about "embodiment"—the embodiment of the mind. The brain does not simply represent the world in a disembodied way as an intellectual construct, but rather the organism interacts as a whole with the environment. In this notion, a mind separate from the body (which according to René Descartes would be an autonomous world) does not exist, but rather our mind is body-bound. We think, feel, and act with our body in the world.[5] All experience is embedded in this body-related being-in-the-world. Or, to put it another way, subjective experience means living that is embodied in the environment and social interaction with other people.

As psychological and neuroscientific studies of the last decade as well Gabriel Revault d'Allonnes's individual case study confirm, temporal experience alters according to the intensity of emotions and is linked to the accompanying physical processes. In this, the experience of self and the sense of time fluctuate in their intensity. In periods of boredom, the perception of self and the sense of time (which is ticking away all too slowly) are both intensified. Before the climax of ecstasy, temporal experience expands and the sense of self is heightened in the extreme. In an extreme experience of flow or even in mind wandering, by contrast, we perceive ourselves and time less intensively. Individual cases such as that of the patient Alexandrine are not representative. But they are the source of new findings about relationships, which must then be systematically tested. In the case of Revault d'Allonnes's work, it took a hundred years before basic psychological and neuroscientific research was able to catch up with the description of Alexandrine's particular case.[6]

In 2009, the brain researcher A. D. (Bud) Craig of the Barrow Neurological Institute in Phoenix, asserted that temporal experience was dependent on body signals.[7] According to Craig, the sense of time passing arises from the continuous construction of representations of the body in the brain. The autonomic neural pathways that connect body and brain run from the lateral gray column of the spinal cord to the brain stem and then via specific thalamic nuclei to the posterior insular cortex, which is part of the cerebral cortex. Through a gradual processing and integration of these body signals with other sense impressions and thought processes and with motivational and situational conditions, an all-encompassing representation of the present condition is created in the anterior insular cortex. This is what Bud Craig calls a "global emotional moment," and it represents the self, updated from moment to moment, as "how I feel now." Consequently, the concept of self is based on physical conditions that alter with time; in this way the sense of time arises as a perception of self through time.

My own research at the University of California in San Diego using functional magnetic resonance imaging (fMRI) generated the first explicitly discussed empirical findings that indicated the involvement of the insula in temporal perception.[8] In this fMRI study, activity in the posterior insular cortex in particular demonstrated a steady climb over time, finishing only at the end of the 9- or 18-second temporal intervals to be estimated by the subjects. During the phase when the test subjects had to press a button to indicate that a certain period of time was now just as long as a previously perceived interval, the anterior insular cortex and sections of the frontal cortex showed this rise in activity; it declined only shortly before the button press. Following Bud Craig's thesis, these results can be interpreted in the following way: through the activity in the insula, the passage of time is coded as an increase in body signals, initially in the posterior insular cortex. When comparing the time interval that is just being perceived with the one presented earlier, the anterior insular cortex, together with sections of the frontal cortex, is particularly active as a conscious temporal comparison is taking place that increasingly involves thought processes, which include working memory and decision-making.

The bodily feelings that are linked to the insula—body temperature, pain, muscular contractions, physical contact, and signals from the gut—are also an integral component of emotions and trigger positive or negative feelings. Short-term affects as well as longer-lasting moods are essential for the modulation of the sense of time. Consequently, physical sensations, emotions, and time consciousness are closely interconnected. As in the case of Alexandrine, these areas can be jointly impaired—whatever the psychiatric or neurological causes may have been. The consciousness of self, as a physical and emotional self, as well as time consciousness, are inextricably linked.

Today's discussion[9] of the connection between self, body, emotion, and time was already advanced in the mid-twentieth century by the French philosopher Maurice Merleau-Ponty in his *Phenomenology of Perception*: "We must understand time as the subject and the subject as time" (p. 490).[10] Merleau-Ponty equated time as a mental entity with a body:

> It is as much of my essence to have a body as it is the future's to
> be the future of a certain present. So that neither scientific thema-

tization nor objective thought can discover a single bodily function strictly independent of existential structures, or conversely a single "spiritual" act which does not rest on a bodily infrastructure. (p. 501)

This refers to all mental functions. In addition, the sense of time is "embodied" in a more all-encompassing way than the other senses. Ultimately, time perception is not mediated by a specific sense organ, as happens in the case of the senses of sight, hearing, taste, smell, or touch. There is no sense organ for time. Subjective time as a sense of self is a physically and emotionally felt wholeness of our entire self through time.

Me, Myself, and I: Boredom, Depression, Meditation

Time and self are one: this is a powerful hypothesis put forward by phenomenology and the neurosciences. If it is correct, then perception of self and time perception would have to be interchangeable. To put it another way, the concept of self and the experience of time should display a high correlation. As became clear in chapter 1, extraordinary states of consciousness entail precisely this. But even in everyday states of consciousness the relationship is clear to see. In situations we find boring, we are forced back on ourselves and feel ourselves in an unpleasantly intensive way. On a boring Sunday afternoon, time spent alone passes only too slowly. Psychological studies show that people who are prone to feeling bored experience time passing more slowly when engaged in monotonous activities than do people who are seldom bored.[11] Boredom actually means that we find ourselves boring. It's the intensive self-reference: we are bored with ourselves. We are tired of ourselves.

In his discussion of the philosophy of Martin Heidegger, Rüdiger Safranski provided an incisive description of boredom:

Pure time, its pure presence. Boredom—that is, the moment when no one notices that time is passing because it will not just then pass, when one cannot drive it away, make it pass, or, as the saying goes, fill it meaningfully. ... It refuses to pass, it stands still, it holds one in inert immobility, it "thralls." This comprehensive paralysis reveals that time is not simply a medium in which we move, but that it is something that we produce out of ourselves.[12]

According to Heidegger, boredom leads crucially to the problem of time, and thus to ourselves, "to an understanding of *how time resonates in the ground of Da-sein.*"[13] As unpleasant as it may be, in a state of boredom and the extreme experience of time associated with it we come closest to ourselves. We are completely ourselves without distraction, in Heidegger's words: "This peculiar impoverishment which sets in with respect to ourselves in this 'it is boring for one' first *brings* the *self* in all its nakedness *to itself* as the self that *is there* and has taken over the being-there of its Da-sein."[14] In boredom we are completely time and completely self—inner emptiness. Now I am I and nothing else—a surfeit of being oneself, in most cases when one is alone, but sometimes also being lonely when being with others.[15]

Depressives complain that time passes terribly slowly and hours can sometimes feel like whole days. "It's only five o'clock and the day is still not over. I wait and wait for evening, and must constantly look at the clock," as one 37-year-old man suffering from depression said.[16] A US study in the 1960s looking at 50 patients with different psychiatric diagnoses, including 20 depressives, who were admitted to the emergency department of a hospital and were tested again on average nine days later as patients in the hospital, showed how emotional symptoms and temporal experience correlate. The greater the improvement in emotional symptoms (fear, aggression, depression) during their hospitalization, the more accurate their estimation of time became, moving in the direction of a relative acceleration.[17]

People suffering from depression are temporally desynchronized; their internal speed does not match the speed of the social environment.[18] Depressiveness and sadness, expressed in a negative self-image, self-blaming, and a feeling of worthlessness, among other things, go hand in hand with the intensified, unpleasant sensation of time passing more slowly.[19] In the absence of a positive future perspective, one's personal situation is experienced as imprisonment in the now of slowly passing time.

But being completely present in the "here and now," as in experiences described in chapter 2, is also the desired condition in meditation. Whereas boredom can creep up on us and take over almost imperceptibly, contemplation in meditation entails the positive will to transcend self and time. However, beginning meditators, in particular, often have to struggle with feelings of boredom as they focus on themselves.

In order to describe situational boredom, we do not need to invoke the notion of existential *ennui*, the inner emptiness of a Sunday afternoon, as actually we could do anything we wanted to. In fact, personally and professionally, we are fine, but nevertheless we feel listless. Our *joie de vivre* evaporates for hours on end. Already in the mundane activity of waiting for the train or bus, or sitting in a traffic jam, we are constantly confronted with ourselves. This is why so many people reach for their smartphones while waiting on the platform or on a five-minute commuter journey. Happily, there's always the radio, if you are stuck in the car in a traffic jam. Psychologists are researching these everyday situations and have discovered that if people have to spend just 6 to 15 minutes alone in a room with themselves and their thoughts, many judge this period of time to be totally unpleasant and are grateful for any diversion, even mildly painful stimuli, just to be distracted.[20] And this applies not only to impulsive young people and adults, but to a representative selection of adults of all ages.

This feeling of an intensified and particularly negative sense of self-awareness goes hand in hand with a specific activation of emotional networks in the brain, as was shown in an fMRI study[21] using test subjects who experienced situations of relative boredom for short periods during a first-person shooter video game. Depending on the extent of decrease in positive emotions, the anterior insular cortex was bilaterally active, along with other important areas that regulate the emotions. In addition, an increase in negative affect was associated with activation of bilateral anterior cingulate cortex. This area is closely linked functionally with the anterior insula. According to Bud Craig's model, the organism's actual state in the moment is generated by the integrated processing of all available information in the anterior insula. This is how I feel *now*. The regulation of behavior, caused by the deviation of an actual state from a target state, is initiated by activity in the anterior cingulate cortex. When it gets too hot (actual state), we look for a shady spot in order to restore the target state of relative coolness. When we are thirsty, we drink. When we are bored, we look for another, more inspiring situation—or take out our smartphone.

Other areas involved in modulating the experience of self and time have also been explored using fMRI studies. As we saw in chapter 2, in mindfulness meditation our attention is focused on the experience of the "here and now"; every moment should be perceived in a concentrated way. To maintain the experience of the present, one directs one's focus toward specific areas of the body; for example, attentively monitoring regular inhalation and exhalation—this is intended to help avoid mind wandering. This procedure leads to an increase in the experience of bodily presence, and thus directly to the experience of time expansion. For example, fMRI studies using trained meditators show how the induced feeling of presence is linked to greater activation of the insular cortex.[22] Research by Ulrich Ott and Britta Hölzel at Gießen University's Bender Institute of Neuroimaging, conducted with the participation of long-term meditators, also showed an increase in insular cortex volume, compared with people who did not meditate.[23] This suggests that experience of the body self, intensified over the years, leads to an enhancement in the formation of connections between the nerve cells in this area of the brain—yet more powerful proof of the brain's plasticity. Other brain areas are active in meditators and are larger in experienced meditators: for example, the hippocampus, which bears a key responsibility for the storage of new experience; and the frontal cortex, which is linked to the executive functions of attention and working memory.[24] As became clear in chapter 2, an increase in mindfulness is linked to a greater capacity for attention regulation; in addition, the working memory is essential for the maintenance of mental presence. Mindfulness means that events are experienced more consciously, leading to increased and more detailed storage of events.

While activity in the insular cortex increases during meditation as a correlate to the increase in physical awareness, at the same time activity in the midline region of the brain decreases.[25] This corresponds very well with the concept of the so-called default mode network, the cortical midline structures. When fMRI was first used to examine brain activity, it came as a complete surprise to some researchers that during the breaks between tasks, when apparently nothing was happening, powerful neuronal activity was registered in the cortical midline structures.

To any test subject who has ever been in the fMRI scanner, however, this will not be particularly surprising. It is precisely during these quiet break periods when mind wandering begins: "What am I doing here?"; "I hope it's over soon"; "What a boring job!" In the absence of stimulation and tasks, test subjects give their thoughts free rein, reviewing past events or looking forward to an evening with friends. And this mind wandering is associated with activity in the default mode network.[26] These are thoughts related to the self—the imagined self of past, future, and possibility. The studies conducted by the medical scientist and philosopher Georg Northoff in particular have shown how this kind of narrative self-referentiality is linked to the cortical midline.[27] For Northoff, the default mode network is the neuronal basis for what William James called the "stream of consciousness," the constant flow of perceptions, thoughts, and fantasies. Many studies using imaging techniques show that the self-referential imagination, producing scenes from the past as well as in the future, is linked to the default mode network.[28]

Of course involuntary mind wandering happens not only while lying in the scanner, but in all periods of waiting. Even during meditation, thoughts slip in unbidden and we lose focus on the present experience. Experienced meditators have demonstrably less activity in the default mode network during a session of mindfulness meditation.[29] This means their heightened mindfulness correlates with a decrease in activity in the neuronal networks linked to mind wandering. Thus we might say that changing between conscious experience of our self in the present moment and having a wandering mind goes hand in hand with a change in predominant activation between the two brain networks of the insular cortex and the cortical midline. In heightened states of presence—associated with insular activity—time slows down, and the felt "now" expands. By contrast, when our mind wanders more—correlating with activity in the cortical midline—time accelerates.[30]

Time passes quickly when we are feeling entertained and when we are absorbed in various kinds of activity. In common parlance we even say we are "lost in our work." We don't quite perceive ourselves; all our attention is directed outward. This extreme loss of self is the experience of flow, which can occur during activity that is particularly intensive but comes easily and smoothly—while writing this book, while playing music, while practicing a hobby. We are totally engrossed, and attention is focused effortlessly on the performance of the activity. We do not perceive ourselves, and as a result subjective time accelerates. Typically, as soon as we become aware of ourselves and of the time, we are astonished at how much time has elapsed. In the flow of intensive activity, the perception of self is reduced and the feeling of time disappears. This is a paradoxical situation. On the one hand we have achieved something that will be permanent—writing this text, solving a syntax problem in programming—but our life as a whole has almost disappeared for minutes or even hours. We were concentrating fully and completely on the matter at hand, but in doing so we did not notice ourselves: a loss of the experience of both self and time. Expressing it negatively this way also shows how the perception of self and that of time are jointly modulated.

Epilepsy and Timelessness

In the extreme case of extraordinary alterations in consciousness, dramatic modulations take place in self-awareness and in the sense of the body and time. A feeling of spatial unity of the body and self with the surrounding space can occur—oceanic boundlessness—or a sense of timelessness can be experienced, as a unity of past and future in the present. Such alterations in consciousness are described in neurology too, for example in the case of certain forms of epilepsy. In these instances, there is initially an excessive feeling of presence. What ultimately occurs is the borderline experience of alteration in physical image and in spatial and temporal experience. In this context, we speak of epileptic auras, which can be tracked using hospital measuring equipment. Studying this neurologic condition, one sees what happens in the brain while extreme alterations in consciousness take place.

During an epileptic seizure, clusters of neurons discharge simultaneously. A characteristic of such focal seizures is that these discharges are confined to particular brain structures. Just before they lose consciousness—which does not necessarily happen; in the case of consciousness disorders and absence seizures, consciousness is sometimes maintained—affected individuals experience special perceptual impressions. These impressions are linked to the mental functions located in the affected regions. For example, in occipital lobe seizures, visual hallucinations can occur if the seizure remains confined to that area, or before the seizure spreads.

The Russian writer Fyodor Mikhailovich Dostoyevsky, who suffered from epilepsy, provided precise descriptions of his ecstatic auras in his novels and letters. Moreover, his wife, Anna Grigoryevna, wrote about her recollections of her observations and conversations with him. For example, the seconds preceding a seizure were often associated with incredible feelings of happiness and of meaning in life which—typically for mystical experiences—were indescribable.[31] After such a seizure,

Dostoyevsky's suffering resembled the physical after-effects of drug intoxication, while mentally he lost the ecstatic state of happiness. Besides these feelings of depression, his memory and attention were functionally impaired for days. In his novel *The Idiot*, Dostoyevsky writes about Prince Myshkin's ecstatic state during auras that lasted for seconds:

> The sense of life, of self-awareness, increased nearly tenfold in these moments, which flashed by like lightning. His mind, his heart were lit up with an extraordinary light; all his agitation, all his doubts, all his worries were as if placated at once, resolved in a sort of sublime tranquillity, filled with serene, harmonious joy, and hope, filled with reason and ultimate cause.[32]

Finally, Dostoyevsky, alias Prince Myshkin, describes the ecstatic auras as an extraordinary state of consciousness:

> Those moments were precisely only an extraordinary intensification of self-awareness—if there was a need to express this condition in a single word—self-awareness and at the same time a self-sense immediate in the highest degree.

By using such terminology, Dostoyevsky describes an extraordinarily intensified feeling of presence.

In the last few years, neurologists have been looking more and more at such experiences. An increasing number of patients have been examined in hospitals using imaging technology in order to locate the area of the brain with the most activity during a seizure. Studies available so far from Geneva and Linköping indicate that the anterior and central insular cortex is particularly activated during ecstatic auras.[33] The patients in these studies convincingly report feelings of being powerfully physically and mentally present and aware of themselves (this recalls states of meditation); how the sense impressions are intensified and that the physical feeling can be compared with an unprecedentedly intense orgasm (this reminds us of accounts of drug intoxication); how they are in harmonious unity with themselves and the world; and that time and space no longer have any meaning (akin to mystical experiences or the effects of hallucinogens). Some patients talk about a transcendent force, sometimes about feeling the presence of a divine power.

First of all, these accounts correlate with observed brain events that indicate the involvement of the insula in the creation of a sentient self. In addition, the phenomena of extraordinary states of consciousness seem to be linked to an uncontrolled overactivation of the insular cortex. A heightened feeling of presence is produced, and this can ultimately result in a dissolution of the boundaries of the bodily self in space and also a dissolution of the sense of time. If confirmed, these findings illuminate how the physical sense of self is modulated by processes taking place at a neuronal level. A sensual exaggeration of the self goes hand in hand with an increase in the activity of the insular cortex. Beyond a certain threshold, if the activity becomes too strong, the senses tip over, leading to the disappearance of the feeling of self.

A study published recently by a group of French researchers has added to the data with the case of a patient presenting epileptic seizures in the right anterior insula.[34] Uncontrollable with medication, these seizures were caused by a slow-growing tumor (astrocytoma) that had been surgically removed. The anterior insula had been damaged by repeated seizures before the patient could undergo an operation. Striking difficulties were observed in the temporal estimation of intervals on the scale of seconds—difficulties not displayed by patients with epileptic seizures in other areas of the brain. Of course, disorders in time perception can initially be regarded as merely secondary when linked to a patient's neurological illness. But in the context of the data, which supports a comprehensive alteration in consciousness in this kind of epilepsy, this record of the experience of time acquires significance as an important marker for self-consciousness.

Schizophrenia, or When Time Stands Still

The psychiatric illness of schizophrenia is a disorder in which the experience of the self is massively impaired. Typical symptoms are delusional mental disturbances (sufferers feel they are being watched and followed) as well as auditory hallucinations (they hear voices making comments). Schizophrenia is primarily a disturbance of the "self," as described by psychiatrists since the beginning of the twentieth century.[35] In cases of schizophrenic psychoses, the phenomenon of self-alienation occurs: the sufferer feels threatened by the dissolution of the self. The "self" is experienced as permeable, the ego boundaries merge with the world, and a terrifying dissolution of the self results—a state that can also be induced by ingesting hallucinogens such as LSD. In addition, emotions, thoughts, and actions are experienced as alien or unreal. The "self" is no longer understood as a unity—it fragments, as it were. This is the account of one female patient:

> I think I am dissolving. I feel—my mind feels—like a sand castle with all the sand sliding away in the receding surf. ... Consciousness gradually loses its coherence. One's center gives way. The center cannot hold. The "me" becomes a haze, and the solid center from which one experiences reality breaks up. ... There is no longer a sturdy vantage point from which to look out, take things in, assess what's happening. No core holds things together, providing the lens through which to see the world.[36]

This description clearly shows how the observer's perception is pathologically impaired. It addresses the subject of perception, or, to put it another way, the processes generated by the subjectivity of experience. In normal circumstances, these processes create a unity of experience that underpins a unitary sense of the self as a spatial and temporal presence in the stream of consciousness.[37] The unity of the self—at the phenomenal level, the self-consciousness—is on the one hand linked to the body as its frame of reference. My perception is perspectival, that is, related to the location of my body with its particular characteristics. On the other hand, I experience myself as someone who is perceiving both now and through time.

As numerous accounts of schizophrenic patients' experiences record,[38] self disorders in schizophrenia are accompanied by disorders in time consciousness. The temporal modes of past, present, and future shift in emphasis. The relationship with the future is experienced as unreal, and as a result the past becomes more intensely present. In a scientific paper published in 1929,[39] the psychiatrist Franz Fischer reproduces the following account by a female patient (case history 7, patient Ku):

> I stand still; I am cast back into the past by the words spoken in the room. ... The present no longer exists, only a backwards reference. ... Does the future exist at all? Earlier I had a future, now it dwindles increasingly. The past is so intrusive, it throws itself across me, dragging me back. ... All this to say that the future does not exist and that I will be cast backwards.

Others among Fischer's female patients emphasize the loss of the passage of time, and their feeling that time is standing still (case history 6, patient Ze):

> The others walk to and fro in the room, but for me time does not pass. The clock works exactly as it did before. ... Time passing and

the clock hands turning are things I can no longer quite imagine. ... What does the future mean to me? One cannot reach it. ... Time stands still; ... this is boring, extended time without end.

Objective time passes, as the patient can see by looking at the clock. But she can no longer feel the passing of time. If time stands still, equally the future can no longer exist.

Another patient, an eloquent philosophy student, also reports feeling that time is standing still. He clearly articulates how difficult it is for him to capture the subjective sense of time in words (case history 8, patient Ge):

Time broke down and stood still. In fact it happened differently, as time appeared as immediately as it disappeared. This new time was infinitely multi-form and intercalated, and in fact hardly to be compared even distantly with what we otherwise call "time." ... Thinking stood still, indeed everything stood still as if nothing existed any more. I seemed to myself to be a timeless creature, extremely clear and transparent in my psychic connections, as if I could see into the very heart of myself.

This patient even tries to override the temporal standstill using rhythmical movements, to produce a continuation of time, as it were:

It was a standstill, a freezing of the inner oscillation that otherwise permeates us psychically and physically and that can be called the "feeling of life." In a nervous rush I attempted to help myself get over this by drumming my fingers on the edge of the bed, but without success. In doing this I attempted to beat out a rhythm, and I still well remember how important the very creation of a rhythmic beat seemed to me. Why, I can no longer recall.

It is striking—and this recalls the similar expansion of time and space in Walter Benjamin's experience of hashish—that even the perception of space has changed markedly. The patient (Ge, case history 8) describes it as follows:

The space seemed to expand, to grow infinitely, and at the same time was as if emptied out, such as the time when I walked across the fields one evening. I felt incomparably alone and abandoned, powerless to fulfill myself and exposed to the infinitely broad space, which despite its emptiness stood threateningly before me. For me it was the immediate extension of my own emptiness and my psychic collapse.

Here the patient experiences the threatening ego disorder in relation to the space around him as isolation. Small and empty, the self faces the great surrounding world.

In schizophrenia, the continuity of temporal experience and with it the continuity of the self are disturbed.[40] It is as if the "self" is stuck in the present. Time no longer moves on, and seems to stand still. Temporal standstill means the standstill of the subject. Normally we experience ourselves as a unity of our self. Our focus on anticipated events kick-starts our preparations for action. Mental presence means that we integrate past, present, and anticipated experience into a whole that is our self. As conscious beings we are constituted through self-experience in the three temporal modes. As the Heidelberg psychiatrist Thomas Fuchs puts it,[41] in schizophrenia the "intentional arc" of our self-consciousness is disordered across the three temporal perspectives. The dynamic of the passing of time, which underlies the subjectivity of all our experience, no longer functions. Because subjective time "gets stuck," the experience of the self that depends on the underlying dynamic temporal structure is impaired. Without the dynamic of this temporal flow, the "self" collapses into fragments of now.

Of course, in such a complex syndrome as schizophrenia, many areas and neuronal networks are affected.[42] In particular, the connections between the brain areas seem to be disturbed. fMRI studies carried out by researchers at Munich's Technical

University were able to establish in detail the altered functional connectivity between three areas in the brains of schizophrenic patients:[43] (1) the default mode network in the midline of the cortex, which—as previously described—is linked to mental "time travel" into the past and the future; (2) the dorsolateral frontal cortex, which is associated with the executive functions of attention and working memory; and (3) the so-called salience network, consisting of the anterior insular cortex and the anterior cingulate cortex. "Salience network" means that these areas are involved in the detection of external, but also and in particular, internal (i.e., emotional and bodily) states.[44] In psychotic patients, the anterior insular cortex showed a conspicuous decrease in activity and in connectivity with other areas. This can be interpreted as an indication that decreased and disturbed bodily perception forms a basis for psychotic illness—with an impact on the immediate experience of the self and of time.

Some researchers, such as Sanneke de Haan and Thomas Fuchs, assume that the main disturbance in the case of schizophrenia lies in the lack of "embodiment," as if the mind is not correctly anchored in the body.[45] They claim that, phenomenologically speaking, the "self" lacks sufficient contact with physical processes, leading to disturbances in awareness of the body and of the emotions, and in temporal experience. This analysis might also explain the disorder presented in the case of Alexandrine, the patient whose story opened this chapter. Further fMRI anomalies in midline activity are highly consistent with the reported disturbances in temporal perspective. In addition, mental presence is dependent on the functioning of attention and working memory processes, which in turn are closely linked to activity in the frontal cortex. While exercising all due care not to overinterpret such results using imaging technology—which can only give an indication of possible associations—the researchers at Munich's Technical University have identified their patients' altered brain function in structures that are also closely associated with time consciousness.[46]

When schizophrenia patients are asked to estimate periods of time in the range of seconds, they indeed present as impaired. Studies carried out over the last few decades show that they frequently overestimate a period of time of several seconds.[47] Short time periods of less than one second, by contrast, are if anything underestimated or temporally discriminated less accurately, probably because attention and short-term memory are not functioning optimally, i.e., the memory traces disappear more quickly.[48] Consistent with the Munich fMRI study, the results of these studies on time perception using short duration indicate an impairment of the executive functions, as has been well documented in the pathology of schizophrenia.[49]

Correspondingly, many patients report time coming to a standstill and the simultaneous loss of the future perspective. These sensations are connected: because the relationship to the anticipated future has been lost, the feeling of time at a standstill arises. Can this subjective experience be recorded experimentally? Can we show at a behavioral level that schizophrenic patients get stuck in the present moment and their relationship to the future is limited? For a long time, subjective experiences of patients were recorded by psychiatrists who were actually interested in people's lived experiences. They were, in particular, psychiatrists who had a philosophical education and were interested in phenomenology, and who wanted to understand the conscious experiences of their patients beyond a purely scientific worldview.[50] To date, however, these reports have hardly ever been used to understand the illness and possible therapeutic approaches. Their shortcoming is that they were at most descriptive, and by no means all patients are eloquent enough to be able to express their complex experiences in words. If there were an objective method of recording subjective time consciousness, one could then examine patients purposefully, follow the course of the illness, and assess the success of therapy.[51]

Perhaps the most promising approach to the objective recording of time consciousness in schizophrenic patients is currently being pursued by Anne Giersch and her team in the psychiatric department at Strasbourg University Hospital. A computerized procedure is used there, which examines the ability to recognize the temporal sequence of visual stimuli. For a test subject the task is notionally simple. On a screen, two squares appear in two locations, on the left and on the right. However, most of the time the squares are shown not simultaneously, but with a short time interval in between.

If the interval lasts for a sufficient length of time, let's say a quarter of a second, a person can easily recognize the temporal order: first right, then left. Eventually, however, as the intervals become increasingly shorter, the ability to recognize the temporal order suddenly and consistently decreases. That is, from a certain interval onward, the accuracy rate becomes increasingly worse. Nevertheless, the person can clearly see that the squares do not appear at the same time. Even if one can no longer clearly make out the exact temporal sequence, one can still see that the squares did not appear simultaneously; one has the impression of visual asynchrony.

In the experiments carried out by Anne Giersch and her colleagues, test subjects were asked to press one of two keys, positioned left and right, to say whether the two squares appeared simultaneously or asynchronously.[52] For example, the left key was to be pressed if the stimuli were simultaneous, and the right key if they were asynchronous. At some point, when the temporal difference between the squares was between 20 and 40 milliseconds, the response was almost exclusively "simultaneous." This temporal fusion of the two events, one could say, represents the *functional moment* discussed in chapter 2. Under these circumstances, the test subjects could no longer recognize the temporal asynchrony. An initial result from a sequence of tests is that schizophrenic patients answered "simultaneous" even in the case of longer time intervals between the stimuli, where healthy participants in the experiment could still clearly recognize "one after the other." From these and many similar experiments, for example using auditory or complex visual/auditory stimuli, the researchers concluded that schizophrenic patients have an extended "now" moment.[53] They need longer intervals between two stimuli in order to experience them as temporally distinct. We can interpret this as the objective correlative of the altered "now" perception in the case of schizophrenic patients—the feeling of time at a standstill and of expanded moments.

But this is not all. In the aforementioned analysis, the issue is the conscious perception of temporal asynchrony, of the explicit indication of temporal sequence. As Anne Giersch's team of researchers in Strasbourg went on to discover, the patients' implicit temporal processing is also striking. This was revealed through the unconscious behavior of the test subjects in the case of very short time intervals. When the temporal distance between the two stimuli is less than 20 milliseconds, it is impossible to consciously perceive the asynchrony. Test participants in that case pressed noticeably more often, although not always, on the "simultaneous" key. But an important effect distinguished between the two groups, even in the case of these short intervals. Healthy people demonstrated an unconscious motor "tendency" to press more frequently on the key that appears on the side of the second stimulus. Even if the test subject perceives the two stimuli as "simultaneous" he or she still shows a higher probability of pressing on the key that is on the side of the second stimulus. This means that in our system of perception, temporal sequence is processed unconsciously, and we just don't notice it.

By contrast, schizophrenic patients displayed the tendency to press more frequently on the key located on the side of the first stimulus.[54] This led to the interpretation that the patients—like the subjective reports—have become stuck in the "now," and that they cannot follow the temporal sequence of the visual stimuli even unconsciously. Consequently, patients with schizophrenia process the stimuli in isolation, and not in a temporal sequence. The astonishing result is that at an implicit level the patients cannot look "forward" into the future, but get stuck in individual moments. This work potentially represents the discovery of an experimental paradigm that records an implicit behavioral disorder, corresponding to the subjective experience of the patients. It is surely only a first step in developing, at an objective level, a measuring procedure that can be used in future studies. It will, of course, be a question of assessing such abnormal timing behaviors in the context of clinical parameters; this is extremely important for the patients' diagnostic and therapeutic support. Nevertheless it is possible that, through these results, research has taken an important step. It has been a long journey from the purely phenomenological recording of subjective symptoms in the disintegration of the self and the altered perception of time to the development of an experimental paradigm.

Timeless through the Doors of Perception

How should we evaluate hallucinogens? Are LSD, psilocybin, mescaline, ayahuasca, and all related substances dangerous intoxicants? Or do these substances provide us with access to another reality, like a vision of the divine? Highly diverse stories are in circulation on this subject, and each in its context identifies at least one aspect of reality. This is the case with alcohol, too. It is the means of overcoming social inhibitions, a brain poison that creates dependence in millions of people, an indispensable cultural asset, or even the blood of Christ.[55]

In the 1950s and 1960s, well over 1,000 clinical studies on the effects of LSD and psilocybin were amassed in medical and psychological journals.[56] For example, psychodynamic experiments were conducted in order to reach the hidden layers of the unconscious by using hallucinogens. Another approach pursued the recording of mystical experiences induced by drugs. The "Good Friday Experiment" of 1962 became legendary.[57] Twenty theology students attended Mass on Good Friday in a separate room in a church. Ninety minutes before the Mass, ten of the students were given psilocybin, and the other ten received a control compound with no hallucinogenic effects. Using a *lege artis* approach, the experiment was administered double blind. Double blind means that when the substances were given, neither the researchers nor the students knew who received psilocybin and who an inactive capsule.

The two-and-a-half-hour-long Mass then began, with organ music, singing, and contemplative prayers. It soon became apparent who had taken psilocybin and who had not. Ten students sat devoutly, but ten students left their seats, lay on the floor, and walked around in wonder, mumbling prayers.

One played strange music on the organ, another adopted the position of Christ on the cross.[58] Immediately after the Mass, the theology students were interviewed about their experiences during the Mass. When the data were deanonymized, it was confirmed that those who had received psilocybin had had significantly more powerful mystical experiences during the Mass.

In the rebellious 1960s, a crazy blending of science and cult occurred: we might remember the psychologist Timothy Leary, who started as a researcher at Harvard but was soon relieved of his teaching duties by the university, gaining the status of the LSD guru of a spiritual liberation movement. The narrative of the era alternated between mystical and psychotherapeutic promises of salvation and the condemnation of highly dangerous substances that might be ingested by anti-establishment hippies.

Today there is a veritable renaissance in hallucinogen research, free from the ideological skirmishing of the 1960s. For example, psilocybin is administered in schizophrenia research, in order to create in healthy participants short-term states resembling acute psychotic phases.[59] Besides this, psilocybin is used in medically controlled sessions, because in the correct setting it can provoke mystical experiences that are potentially capable of altering the personality.[60] In general, a person's characteristics change only moderately once they have reached adulthood.[61] A study led by Roland Griffiths of Johns Hopkins University in Baltimore showed that even a year after the supervised administration of psilocybin, positive changes in the personality trait of "openness" could be seen. The more powerful the mystical experience had been, the stronger the characteristic of "openness to new experiences." In further studies with terminally ill patients, it could be shown that administration of psilocybin alleviated the fear of death.[62] High doses of psilocybin led to marked decreases in depressed mood and death anxiety. At the same time optimism and the sense of life's meaning increased considerably. Since participants of the study reported typical mystical-type experiences, these positive changes may be attributable to a stronger feeling of connection with the world and with the people around them. Even today, the cultural significance of hallucinogens varies according to the different scientific perspectives and stories that circulate about them. Narratives range widely, from the biological reductionist approach of the neurosciences, via holistic approaches (in psychotherapy and palliative medicine), to emphasis on spirituality and mysticism in descriptions of the immediate experience and the associated interpretations.[63] For example, psilocybin-containing mushrooms, which occur throughout the world, or ayahuasca, a brew derived from a variety of Amazonian plants, are taken during shamanistic and religious ceremonies.

Scientific research on the effects of LSD and psilocybin has shown clearly that the states of consciousness involve striking changes in perception, emotions, and ideas, and also in the ways they are described: time, space, and the experience of self are dramatically altered. These changes are comparable only with other extreme states of consciousness such as occur in dreams, in mystical and religious ecstasy, or in acute psychotic phases in the early stage of schizophrenia.[64] The dimensions of mystical experience include oneness of the self with the universe, the feeling of timelessness and spacelessness, the most intense feelings of happiness, and the certainty of experiencing a sacred truth which is, however, indescribable.[65] The latter is the feeling of looking behind the veil of reality and seeing the immutable (that is, timeless and spaceless) truth of the world in its entirety.

This experience also came to Tilmann Lhündrup Borghardt over many years of meditation.

Zurich has seen large-scale pharmacological, neuroscientific, and psychological research into psychoactive substances such as psilocybin, ketamine, and MDMA (Ecstasy) since the 1990s. The doctor who has for decades been leading this research at Zurich University's psychiatric hospital, the Burghölzli, is Franz Vollenweider. Based on a questionnaire devised in the 1970s by Adolf Dittrich in Zurich and developed over the decades to record "abnormal states in waking consciousness," tools were developed that can capture the extraordinary experiences that take place under the influence of psilocybin. Once the effects of the substances have largely abated, study participants are asked appropriate questions. The items on questionnaire 5D-ABZ ("five dimensions of extraordinary states of consciousness") are organized according to the following factors: (1) oceanic boundlessness, (2) dread of ego dissolution, (3) visionary restructuralization (4) auditory alterations, and (5) vigilance reduction.[66] Altered states of consciousness can be described as individual expressions of these dimensions. Factor 3 comprises visual hallucinations, a heightened visual imagination, and vivid memories; in factor 4, changes in auditory experiences are recorded; and factor 5 refers to reduced alertness and a heightened feeling of drowsiness.

In the context of this book, the first two factors, comprising the principal positively and negatively experienced aspects, are particularly important. Over the course of several questions, the "oceanic boundlessness" segment records alterations in the personal experience of self, a perception that is accompanied by an excessively euphoric feeling of elation. One question in particular addresses alterations in the experience of time. This set of questions generally depict the typical, mystical experience in positive terms. But it can be otherwise; the other side of the experience is addressed by factor 2, the dread of ego dissolution. Questions on this topic cover the feeling of losing control over events and the body, the angst-ridden sense that everything is completely unreal, violent thought disturbances, and the nightmarish version of a loss of boundaries between the self and the world. This is a "bad trip." These two factors are experienced in similar ways, though one is perceived as ecstatic and heavenly, while the other is hell.

Over the decades, Franz Vollenweider and his colleagues have only rarely monitored people who had exceptionally unpleasant psilocybin experiences. This could be because prospective test subjects are questioned intensively as to whether they themselves or their relatives have psychiatric disorders or are conspicuously excessive consumers of drugs. Only if there is no personal or family history of psychiatric complications are people allowed to take part in a study. In addition, research participants are monitored the entire time. In an extensive, retrospective study by Erich Studerus, Michael Kometer, Felix Hasler, and Frank Vollenweider, researchers analyzed the experiences of all 110 of their test subjects, who had been tested with different doses of psilocybin in Zurich between 1999 and 2008.[67] The "dread of ego dissolution" factor presented strongly only with the two highest doses of psilocybin used, that is, in around 6.5% of participants who received these highest doses. In the case of one participant, a 23-year-old medical student, longer-term disturbances of some weeks occurred, combined with anxiety and depression that had to be treated with psychotherapy sessions. The test subject reported having been overwhelmed with intensely negative memories while under the influence of psilocybin. In psychodynamic terms, hidden layers had been released, which rose to the surface of consciousness. Indeed, the administration of psilocybin intensifies the vividness of memories and visual images,[68] a fact that could be used under certain circumstances in controlled psychotherapy sessions in the future. However, this particular case also demonstrates that hallucinogens can produce effects that require medical and psychological supervision.

Since the experience of time can be modulated powerfully under the influence of hallucinogens, it makes sense to study the phenomenon in more detail. After taking hallucinogens, people often report extraordinary states of consciousness, strongly resembling phenomena by now familiar to us, such as time appearing to stand still or minutes seeming to last for hours.[69] Can these alterations in subjective time perception also be analyzed objectively as deviations from normal estimations of specified time intervals? In 2003, I was working in the Generation Research Program at Munich's Ludwig-Maximilian University in Bad Tölz. My team, had developed a battery of computerized time perception tasks, which we applied to different groups of subjects. With a computer loaded with these tests I drove to see Franz Vollenweider and his colleagues at the Burghölzli in Zurich, where a major new study was in the process of being designed.[70] A range of procedures for measuring time perception and the temporal control of motor skills was being integrated into the study. It was the first-ever systematic attempt to describe time perception under the influence of psilocybin using objective measuring devices.[71]

The test subjects, who attended the university hospital's consulting rooms on three different days at intervals of two weeks, were given a placebo, a medium dose of psilocybin, and a high

dose of psilocybin, in random order.[72] Each day, they were subjected to time tests before the substance took effect, at the peak
of its impact after 90 minutes, and as the impact diminished
after 4 hours. Those taking the stronger dose presented stronger effects, and at peak impact the effects were greatest: when
asked to temporally reproduce specified sound intervals, the test
subjects who had been given the high dose produced a shorter
temporal duration, deviated more markedly from the objective
time, and were less precise. This effect was, however, only evident in the case of temporal durations longer than 4 seconds;
shorter durations were not affected. When subjects were asked
to rhythmically accompany a regular sequence of sounds, more
mistakes and discrepancies crept in with intervals of 4 seconds,
but not in the case of shorter intervals. The individual tapping
tempo, wherein a self-selected tempo was to be tapped on a button repeatedly and at an equal rate, was also slowed down. This
was, however, not as a result of a general deceleration of motor
skills, as the test subjects were not slower when tapping a maximum tempo. This suggests that the conscious control of the self-
selected tapping tempo was affected.

Are these results explained by the fact that massive alterations in experience and behavior are observed under the influence of psychoactive substances? What conclusions can indeed
be drawn if test subjects tap the button a little earlier during
the temporal reproduction of intervals test? It could be argued
that these are extremely artificial results. But that objection is
countered by the continual increase in effects as the dose rises,
as well as the decrease in impact between measurements taken
at the peak of the effect (after 90 minutes) and measurements
recorded as the effect dwindles (after 4 hours). However, the crucial indication of psilocybin's specific effect on time perception

was provided by a correlation between the objective measure of temporal reproduction and the questionnaire for recording the subjective state, the 5D-ABZ. Indeed, the item "change in subjective time" correlated with the impairment in the timing of the temporal intervals. This means that the subjective feeling of an alteration in the passing of time was modulated together with the ability to estimate duration.

How does psilocybin affect the brain? Psilocybin activates special receptors (called $5\text{-}HT_{2A}$) of nerve cells, which are normally activated by serotonin, a neurotransmitter that occurs naturally in the brain and is closely involved in the regulation of emotions. When hallucinogens are ingested, the activity of the $5\text{-}HT_{2A}$ receptors is massively affected. Using positron emission tomography (PET scan), the research carried out in Zurich on the effects of psilocybin on the brain showed a considerable increase in activity in the frontal areas of the cerebral cortex, including the anterior cingulate cortex and the anterior insula.[73] And what is the link between these alterations in metabolic activity in the brain and conscious experience? In a study carried out in Zurich, the feeling of oceanic boundlessness correlated with an increase in activity in the frontal cortex in subjects who had undergone a PET scan.[74] As further investigations suggest, the powerful activity of the serotonin receptors activates nerve cells in the frontal lobe that release the neurotransmitter glutamate, which in turn activate NMDA receptors. This means that psilocybin has an effect on the glutamate system of the brain via serotonin receptors. These are interesting findings, and they can be related to an idea put forward by the neurobiologist Hans Flohr on the subject of phenomenal consciousness. The activity of the NMDA receptors in complex groups of nerve cells is modulated when certain anesthetics, such as ketamine, are administered:

this effect is important for monitoring unconsciousness during operations, for example. Thus psilocybin indirectly influences the function of a neurotransmitter system that is closely linked to the modulation of consciousness.[75] The massive hyperactivity in frontal areas of the cerebral cortex can be clearly linked to reports of sensory hyperstimulation—of a hyperpresent world in which the self (experienced negatively) threatens to disappear or (experienced positively) becomes one with the whole.

The effects on the brain's neurotransmitters that we have described represent only a selection of all the alterations that have been recorded: for example, the dopamine system is also affected. More recent findings by researchers at Imperial College in London initially created some confusion: they were able to show using fMRI how a medium-sized dose of psilocybin, administered intravenously, led to an immediate *reduction* in activity in many areas of the brain. The research team, led by Robin Carhart-Harris, interprets their results to mean that psilocybin leads to a collapse of normally highly organized brain activity.[76] This disintegration of activity across the brain thus coincides with the frequently experienced disintegration of the self. In particular, reduced activity in the cingulate cortex and a diminished functional connectivity between these regions (the default mode network of the cortical midline, an area discussed earlier in this chapter) and other areas of the brain were related to subjective alterations in the experience of the self. The default mode network is linked to mental "time travel" into the past and the future and is responsible for situating the self within time. Accordingly, disintegration of the self and time and the disintegration of the connections with the default mode network go hand in hand. This shows a correspondence, at least superficially, between conscious experience and underlying neuronal processes.

In these diverse findings, we can see that our knowledge about the effects of psilocybin is still inadequate. In addition, we can see how various methods lead to different results. For example, in the PET scan, the use of a radioactive sugar tracer is measured, while in fMRI the relative oxygen content of the blood is assessed. Also, the researchers in Zurich used capsules, which when taken orally produced their maximum effect only after about 90 minutes. The English researchers, by contrast, administered the substance intravenously, a procedure that is fully effective after just a few minutes. The discrepancy between the findings of the Zurich and London teams can probably be explained by the different means of psilocybin intake.[77] It is also due to the mathematical methods used to derive meaningful results from the data. Using a special kind of data analysis, the London team recently showed that after the intravenous administration of psilocybin, lots of short-term functional connections between areas are established; these were not detected in the case of a placebo.[78] This dramatic increase in fine, functional networks is highly consistent with the massive alterations in perception and mental associations that characterize the effects of psilocybin.

Hallucinogens have a profound effect on both the sense of self and the sense of time. Given the right context, these altered states of consciousness can be of the mystical kind. That's why they are part of religious rituals for indigenous cultures but are also part of philosophical discourse in Western culture. In recent years they have also become a therapeutic means to treat depression and anxiety disorders—in patients with life-threatening cancer, but also in patients for whom standard pharmacological therapies failed. These drugs are also used to treat addiction to alcohol, cocaine, and tobacco.[79] The available studies to date are still considered experimental, since large clinical trials have yet to be conducted. The first results are very promising and show impressively positive effects. In explaining the sometimes astounding transformation an individual goes through, the researchers suggest that the psychedelic "spiritual awakening" gives a person a different perspective on life and a sense of meaning. This could happen through the mystical connecting of the individual self with the world. Similar to what highly experienced meditators report, meaning and happiness grows through this spiritual connection with our fellow sentient beings.

Indeed, hallucinogens can be personality-changing and life-changing, as controlled studies have been able to show. However, as reports on psilocybin's functional mechanisms make clear, the neurobiological mechanisms of psychedelic substances are poorly understood. While rare and spontaneously occurring mystical experiences are hard to explore scientifically, in hallucinogens we have a class of substances that can be used in a clinically controlled way. This opens a window for researchers to investigate extraordinary experiences that people throughout history have undergone. Research into the mystical experience of the disintegration of time and the self under the influence of hallucinogens is a way toward understanding human consciousness.

Epilogue: On Scientific Awakening

Through greater or lesser variations in consciousness we can learn a lot about the processes that underpin the self-conscious mind. However, for a long time extraordinary consciousness experiences have either been ignored by the mainstream natural sciences or have been explicitly denigrated as nonexistent— as the fantasies of cranks.[1] In states of consciousness that are produced during spontaneous mystical experiences for some people, in experiences associated with meditative states, in drug experiences, in states of hypnosis or trance, and sometimes while listening to music, the perception of time and space is altered in extreme ways, as this book has documented. In addition, notwithstanding the variability of such experiences and their circumstances, the accounts of people who have had extreme experiences, whether in situations of danger (near-death experiences) or in the face of particular neurological and psychiatric conditions, further suggest extraordinary states of consciousness wherein time, space, and the experience of self are intensely altered. These phenomena, too, were largely ignored by scientists in the second half of the twentieth century because conventional scientific theories could not furnish any explanation. For example, the ecstatic auras of Fyodor Dostoyevsky that preceded his epileptic attacks, which were associated with incredible feelings of happiness and harmony with himself and the world, have only recently become the subject of serious neurological research.[2]

Times are changing. The very fact that in the past few decades the theme of consciousness itself has become a central topic for psychologists and neuroscientists signals a transformation in the scientific landscape. For example, it is said among distinguished brain researchers that just twenty years ago they dared not disclose that their actual research topic was consciousness—they had to say they were studying visual perception, or processes of attention. The prevailing advice was not to "out" yourself as a consciousness researcher before achieving permanent tenure of employment. Of course by then, psychology was no longer dominated by the kind of behaviorism that, with its stimulus-response models, rejected consciousness as superfluous. But even subsequent paradigms in the cognitive sciences had no need for the concept of consciousness. How much harder it was then for scientists who were researching "altered states of consciousness." If consciousness itself was unworthy of attention, there was certainly no need to investigate altered states of consciousness.

Sociologists of science must take a close look at what has caused this turn toward an acceptance of the phenomena. There is, for example, the punishment of Ignaz Semmelweis in the nineteenth century, whose hygiene rules for hospitals were dismissed as "speculative nonsense," although they merely dictated that after dissecting corpses doctors should wash their hands before attending births on the maternity wards. The Austro-Hungarian doctor Semmelweis had shown empirically that on wards where the doctors washed their hands, cases of puerperal fever decreased significantly. Semmelweis's available empirical findings were simply ignored. A new generation of doctors would have to replace the old before the deadly puerperal fever could be combated effectively.[3]

Had it been introduced just a few years earlier, research by the neurologist Olaf Blanke into out-of-body experiences would have been dismissed as "speculative nonsense." Under experimental conditions, Blanke triggered an out-of-body experience in a female neurological patient at the University Hospital in Geneva. In 2002 he published his findings in *Nature*, one of the two most important natural science journals.[4] During spontaneous out-of-body experiences, people have the impression that they can leave their body and float above it, observing themselves from above. Accident victims also report such experiences, as have some patients after a cardiac arrest (as can be read in the impressive accounts of Péter Nádas) who, to all appearances, were unconscious at the time of the out-of-body experience but who nevertheless can later recount their experience in detail. Patients report having "seen" the scene of the accident, and the procedures carried out by the paramedics and emergency doctors on their body, as it lay on the ground and they observed all this from above.

The patient described by Olaf Blanke and his colleagues had had electrodes implanted to evaluate her epileptic attacks by measuring the seizure activity in response to local stimulation, a procedure in which the patient remains fully conscious. Electrical stimulation of the angular gyrus in the right side of the brain, the temporoparietal junction (and only stimulation of that area), led to selective perceptions of alterations in physical position. In the case of two stimulations, the patient reported a feeling of weightlessness and that she was around two meters above the bed, very close to the ceiling. The patient had no knowledge of which area had just been stimulated. These somatosensory experiences, of a type reported by people again and again, could thus be triggered experimentally and linked to processes in a narrowly circumscribed area of the brain. Of course this does not mean that out-of-body experiences are situated strictly locally in the right angular gyrus; but it suggests that this part of the brain is a significant factor in the occurrence of such experiences, in a way we do not completely understand.[5]

Moreover, some of the stimulations produced the feeling of falling suddenly from a height. Many people are familiar with this sensation, as it sometimes happens shortly before falling asleep: you have the feeling of falling; you might even think you will fall out of bed onto the floor. But nothing has happened— you are still lying peacefully in bed. Olaf Blanke's neurological research also provides clues to these better-known states of consciousness. This feeling of falling at the onset of sleep might also be traced to neural processes in the area of the angular gyrus, operating with some minor glitches.[6]

This transformation in the sciences is also evident in contemporary research into the short-term and long-term effects of meditation, which we discussed extensively in chapter 2. The practices of focused silence and contemplative prayer have existed for millennia. It is striking that over the last decade psychologists and brain researchers have more frequently addressed meditation as a research topic, and this work increasingly garners media attention. Meditation, as a form of psychological intervention, has even been incorporated into hospitals' clinical practice. Research findings on the effects of meditation are clear:[7] just a few weeks' practice of meditation improves performance in terms of attentiveness and short-term memory, effects that demonstrably accompany alterations in brain structure. It can also be seen that those who meditate perceive their bodies more intensively, achieve stronger emotional self-control, and in the long term experience more positive emotions. More important, however, in the context of this book, is the fact that people who meditate regularly have extraordinary consciousness experiences—of calmness of thought, or of the self being one with the world—during their meditation, sometimes even after a short time of practice; more frequently, however, after they have been meditating intensively for many years. For some years these experiences have also been recorded scientifically. For example, in a study that may call to mind disconcerting images, Carmelite nuns were examined using a brain scanner as they prayed, in order to identify the brain activity associated with a mystical communion with God.[8]

But what has enabled such a shift in scientific approach, whereby these "speculative" phenomena more and more are being examined by mainstream researchers? Perhaps it is thanks to a new generation of researchers, who completed their education in a society—reflected in its universities—that is moving toward greater social and psychological openness. Perhaps linked to this new openness is the fact that more people who have had such intensive experiences have by now pursued scientific careers. The neurophilosopher Thomas Metzinger, who conducts research alongside Olaf Blanke, theoretically processing the results pertaining to the relationship between body and consciousness, had a few out-of-body experiences himself as a young adult—experiences that led to a search for answers.[9]

Further, there are scientists whose own formative experiences, whether with meditation or with hallucinogens, influenced their choice of profession; some have gone on to investigate extraordinary states of consciousness.[10] Of course, the employment of established methods of brain research in this work has been a factor in its growing acceptance in the scientific community as a whole. Ultimately, though, it is up to the researchers to first tackle the topic itself, then hold their ground in the arena of more traditional ideas and among entrenched academics who may not be so open to new ones.[11]

"Openness to experience" is one of the five dominant human personality traits (the so-called Big Five). In his book *Consciousness beyond Life*, Pim van Lommel shows how personality influences scientists to ignore phenomena, describing doctors' attitude of denial toward the topic of near-death experience.[12] The book recounts an episode that took place at a conference on the topic, when a doctor responded to a speaker:

> "I've worked as a cardiologist for twenty-five years now, and I've never come across such absurd stories in my practice. I think this is complete nonsense; I don't believe a word of it." Whereupon another man stood up and said, "I'm one of your patients. A couple of years ago I survived a cardiac arrest and had an NDE, and you would be the last person I'd ever tell."

Scientific dogmas and attitudes to life are scarcely ever changed through rational argument. Such shifts require life-changing experiences, of the "road to Damascus" kind, that affect the researcher emotionally. Toward the end of his life and after a heart attack, the logical positivist A. J. Ayer, whose philosophy might be summed up as "only what can be empirically proven and is factually and logically correct exists," had a near-death experience that at least got him thinking.[13] The London *Sunday Telegraph* gave Ayer's description of his near-death experience the strapline "What I saw when I was dead." If Ayer had previously been an avowed atheist, who assumed that there was nothing after an individual's death, he now spoke more cautiously. He still remained an outstanding analyst, and reasoned that for a time after the cardiac arrest brain functions might still exist that could have produced these experiences. In his last essay, "Postscript to a Postmortem," Ayer describes how, affected by his experience, he relinquished his polemical position against the belief in life after death in favor of a still skeptical but nevertheless more open-minded attitude.[14] For him the idea of an afterlife was now at least worthy of research. And indeed for some years there has been vigorous research activity into near-death experience, which is published in the most important medical journals, as we outlined in chapter 1.[15]

Even perhaps the most important living sleep researcher, it is said, had to undergo his own dramatic experience before he began to take dreams seriously and ascribe meaning to them. It is hard to believe, but Allan Hobson, the sleep researcher, discovered only toward the end of his career that "dreams are not just froth."[16] For his entire life as a researcher, Hobson had fought with great conviction and polemic against Sigmund Freud's dream theory, repeatedly emphasizing the random and chaotic nature of dream contents. Then the accident happened: after a stroke affecting the brain stem, in a specific region that is linked to the sleep-wake cycle and which he himself had studied over the course of his career, Hobson was unable to sleep for eight days and did not dream for a month. This brought about a psychotic state in which he began to hallucinate the most incredible narratives. To his annoyance, however, no doctor was interested in the content of his "merely" subjective hallucinations. This life-changing experience motivated him to get involved more intensively and systematically with the subjective aspect of sleep and the substance of sleep consciousness. Only his own perspective was enough to convince him.

If we want to understand our consciousness—our subjectivity —then we must put aside our prejudices and transcend certain boundaries of our own creation. The American philosopher Thomas Nagel describes this attitude of mind in the introduction to his book *The View from Nowhere*:

> I believe that the methods needed to understand ourselves do not yet exist. So this book contains a great deal of speculation about the world and how we fit into it. Some of it will seem wild, but the world is a strange place, and nothing but radical speculation gives us a hope of coming up with any candidates for the truth.[17]

I do not consider the ideas in my book to be radical speculation. I have brought together empirical findings from various branches of the sciences to form a whole. From this emerges a clear picture of the psychological and neuronal foundations of our time consciousness, as it is linked to our consciousness of self. From speculation we reach hypotheses. And, in the empirical sciences, hypotheses are there to be tested. This is the business of the researcher. To be continued.

Acknowledgments

As researchers we are not alone. In order to pursue certain questions, to investigate unusual phenomena, or to interview particular patients, we need enthusiastic colleagues to collaborate with. These research colleagues complement our own position with their years of experience as well as their special methods and individual opinions. The importance of these associations is not just about viewpoints and knowledge, but always about very personal interaction and the willingness to try new things as well. We are richly rewarded by research work with sympathetic, like-minded people.

I am very lucky in my collaborations, for example with Stefan Schmidt of the Department of Psychosomatic Medicine at the University Medical Center Freiburg. With his team we are investigating the influence of meditation on time perception, as well as the mechanisms of voluntary motor skills, linking them conceptually with the question of free will. In our work with Tilmann Lhündrup Borghardt, we collaborate with a meditation master who can exactly describe the subjective viewpoint during meditation and when performing voluntary actions. I am delighted to be able to conduct research with Anne Giersch from the University Hospital of Strasbourg and her motivated colleagues on the topics of mindfulness and meditation, but also especially on schizophrenia. In this endeavor, the recording of patients' descriptions of phenomena involved in time perception is combined with objective ways of measuring temporal experience, as described in chapter 3. I have joined Franz Vollenweider and his colleagues Lukasz Smigielski, Kathrin Preller, and Michael Kometer from the University of Zurich in thinking about some exciting studies on alterations in the perception of time, space, and the self under the influence of psilocybin.

Crucial to the success of our own research is winning third-party funding. The BIAL Foundation (Fundação BIAL in Porto) supported me, Stefan Schmidt, Han-Gue Jo, and Karin Meissner in two experimental studies with experienced meditators. In this way, the already successful collaboration with Karin Meissner of the Institute of Medical Psychology in Munich on the physiological correlates of time perception can be pursued through a dual-center study in both Freiburg and Munich. The doctoral students Simone Otten, Eva Schötz, and Damisela Linares Gutierrez, as well as student assistants Anna Sarikaya, Hanna Lehnen, Ursula Nothdurft, and Sebastian Kübel, should be mentioned here for their commitment and their creativity.

Over the past few years, my lecture tours to conferences and workshops have been financed above all by TIMELY, the European Union–funded association of researchers working in the area of time, in particular ISCH COST Action TD0904, "Time In MEntaL activitY: Theoretical, behavioral, bioimaging, and clinical perspectives." Without this support I would not have been able to present my work internationally and would have missed out on exchanges with wonderful professional colleagues. Through TIMELY's activities, I and many others have come to know and value Wolfgang Tschacher, Claudia Bergomi,

Helena Sgouramani, Mark Elliott, Sean Power, Cintia Retz Lucci, Zhuanghua "Strongway" Shi, Ian Phillips, Christoph Hoerl, Hedderik van Rijn, Martin Riemer, Sven Thönes, Justin Kiverstein, and Valtteri Arstila. I should mention in particular Argiro Vatakis, coordinator of the Cognitive Systems Research Institute in Athens, who brought the group into being and has supported it tirelessly.

I have discussed the themes of this book with many other colleagues, including Niko Kohls, Ulrich Ott, Harald Walach, Wolfgang Achtner, Jannis Wernery, Jirí Wackermann, Kai Vogeley, David Vogel, Thilo Hinterberger, Jürgen Kornmeier, Harald Atmanspacher, Sylvie Droit-Volet, Yuliya Zaytseva, Yan Bao, Ernst Pöppel, Martin Offenbächer, Eric Pfeifer, Carlos Montemayor, Mauro Dorato, Heinrich Paes, Olga Pollatos, Tanja Vollmer, Gemma Koppen, Tobias Esch, Anja Weber, Vanessa Deinzer, Liam Clancy, Rui M. Costa, Julia Mossbridge, Aviva Berkovich-Ohana, Tijana Jokic, Dan Zakay, Oleksiy Polunin, Virginie van Wassenhove, Bud Craig, Justin Feinstein, Sahib Khalsa, and Martin Paulus. In addition, at the Institute for Frontier Areas of Psychology and Mental Health (IGPP), which was founded in 1950 by Hans Bender and where I am employed, I am learning a lot about extraordinary experiences. With four colleagues there, I probably discuss reports of such experiences most frequently. They are Eberhard Bauer, Gerhard Meyer, Ina Schmied-Knittel, and Uwe Schellinger. I have also profited enormously from my contact with many other colleagues at the institute. Thanks to all at the IGPP.

Lastly, my thanks go to all those friends and colleagues who have read and commented on parts of the book. They are Lukasz Smigielski, Barbara Herzberger, Stefan Schmidt, Katharina Weikl, Tilmann Lhündrup Borghardt, Dirk Thiel, Michael Kometer,

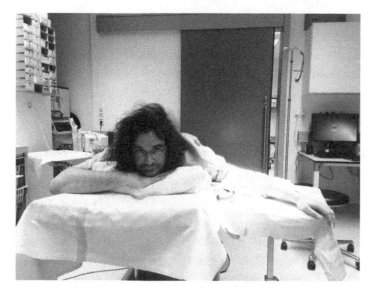

Figure 5.1
An unscheduled visit to the emergency room of University Hospital Freiburg after investigating the effects of a floating tank on consciousness.

Felix Hasler, Johannes Angenvoort, Klaus Meffert, and Oksana Gutina.

That the experience of altered states of consciousness is not without danger was brought home by my experience in a floating tank I visited in Freiburg. Surrendering to my researcher's curiosity, I booked 45 minutes in a floating tank (also known as the Samadhi Floating Tank). In the tank you float, suspended in body-temperature salt water; you can see nothing; and can hardly hear or smell anything. You are alone with yourself and your physical sensations. Astonishingly, the smallest movement is experienced as a major displacement. You travel—weightlessly, so it feels to you—into the dark universe beyond. I like to call this "instant meditation" because after a while you can get into states of consciousness that, normally, only experienced meditators have: there is a diminished sense of an individuated self and a feeling of oneness with the surroundings. Justin Feinstein is the director of the LIBR Float Clinic and Research Center in Tulsa, Oklahoma. He has begun testing the efficacy of repeated floating exposure on people with and without mental disorders. Individuals may feel that they lose some of their self-centeredness and their anxieties. I am particularly happy about this cooperative research with Justin.

In Freiburg, however, gravity got me in the end (figure 5.1). Still completely under the effects of my experience as I stepped out of the shower afterward, I slipped, fell to earth, and struck a sharp edge, sustaining a deep cut to my elbow. And so I experienced, physically and painfully, the earthly foundation of my consciousness.

Notes

Prologue: An "I" Awakes

1. Tomas Tranströmer, *The Great Enigma: New Collected Poems*, trans. Robin Fulton (New York: New Directions Books, 2006), 99.

2. René Descartes, *Meditations on First Philosophy with Selections from the Objections and Replies*, trans. Michael Moriarty (Oxford: OUP, 2008). On page 18, Descartes recounts the idea of a powerful deceiver. Nevertheless, there exists an ego that is necessarily true: "But there is some deceiver or other, supremely powerful and cunning, who is deliberately deceiving me all the time.—Beyond doubt then, I also exist, if he is deceiving me; and he can deceive me all he likes, but he will never bring it about that I should be nothing as long as I think I am something. So that, having weighed all these considerations sufficiently and more than sufficiently, I can finally decide that this proposition, 'I am, I exist,' whenever it is uttered by me, or conceived in the mind, is necessarily true."

3. Marcel Proust, *In Search of Lost Time*, vol. 1, *Swann's Way*, trans. C. K. Scott Moncrieff and Terence Kilmartin, rev. D. J. Enright (London: Vintage, 2005), 4.

4. For the difference between "narrative self" and "minimal self" as well as further conceptual variations on the self, see S. Gallagher, "A Pattern Theory of Self," *Frontiers in Human Neuroscience* 7 (2013): 443. The neurologist Antonio Damasio prefers the concept of "core self" rather than

"minimal self": A. Damasio, *Self Comes to Mind: Constructing the Conscious Brain* (New York: Pantheon, 2010).

5. The minimal self might be understood as a "first-person perspective." Stripped of memory, subjectivity reveals itself directly. It is the subject of experience. In this, subjectivity is constituted as an "ego-pole": I experience myself through the focus of my consciousness upon an object (what has been experienced, imagined, or remembered). But I am also that which experiences or remembers something. The pole consists of the ego that experiences (the ego-subject). However, at the same time the ego is also the experience and the autobiographical memory (the ego-object). But on waking up, the ego-object can sometimes be absent. The minimal self could thus be interpreted as the ego-subject that is thrown back upon itself (self-referentially), while at the same time being the ego-object.

6. Thomas Bernhard, "Walking," in *Three Novellas*, trans. Peter Jansen and Kenneth J. Northcott (Chicago: University of Chicago Press, 2003), 165.

7. On this subject the philosophical examples and empirical research of Jennifer Windt are of particular importance: J. M. Windt, *Dreaming: A Conceptual Framework for Philosophy of Mind and Empirical Research* (Cambridge, MA: MIT Press, 2015); J. M. Windt and V. Noreika, "How to Integrate Dreaming into a General Theory of Consciousness: A Critical Review of Existing Positions and Suggestions for Future Research," *Consciousness and Cognition* 20 (2011): 1091–1107.

8. See my book on temporal perception: M. Wittmann, *Felt Time: The Science of How We Perceive Time*, trans. Erik Butler (Cambridge, MA: MIT Press, 2016).

Chapter 1

1. The notion of "flow" was popularized by the work of the American psychologist of Hungarian origin Mihaly Csikszentmihalyi. M. Csikszentmihalyi, *Flow: The Psychology of Optimal Experience* (New York: Harper and Row, 1990).

2. The relationship between physical experience and the perception of time is developed and discussed extensively in my previous book: M. Wittmann, *Felt Time: The Science of How We Perceive Time*, trans. Erik Butler (Cambridge, MA: MIT Press, 2016). The scientific findings presented here can be found in M. Wittmann, "The Inner Sense of Time: How the Brain Creates a Representation of Duration," in *Nature Reviews Neuroscience* 14 (2013): 217–223.

3. See also M. Wittmann, "Modulations of the Experience of Self and Time," *Consciousness and Cognition* 38 (2015): 172–181; D. Vaitl, N. Birbaumer, J. Gruzelier, G. A. Jamieson, B. Kotchoubey, A. Kübler, D. Lehmann, W. H. R. Miltner, U. Ott, P. Pütz, G. Sammer, I. Strauch, U. Strehl, J. Wackermann, and T. Weiss, "Psychobiology of Altered States of Consciousness," *Psychological Bulletin* 131 (2005): 98–127.

4. Karl Popper and John C. Eccles, *The Self and Its Brain: An Argument for Interactionism* (New York: Routledge, 1983), 529–530.

5. Tomas Tranströmer, *The Great Enigma: New Collected Poems*, trans. Robin Fulton (New York: New Directions Books, 2006), 82.

6. R. Noyes and R. Kletti, "Depersonalization in Response to Life-Threatening Danger," *Comprehensive Psychiatry* 18 (1977): 375–384.

7. A. Heim, "Notizen über den Tod durch Absturz," *Jahrbuch des Schweizer Alpenclub* 27 (1892): 327–337.

8. V. Arstila, "Time Slows Down during Accidents," *Frontiers in Psychology* 3, no. 196 (2012).

9. Psychological research in recent years has shown quantitatively through experiments how time duration is overestimated in emotional contexts relative to neutral situations: S. Droit-Volet, "Time Perception, Emotions and Mood Disorders," *Journal of Physiology* (Paris) 107 (2013): 255–264; M. Wittmann, "The Inner Sense of Time," *Philosophical Transactions of the Royal Society B* 364 (2009): 1955–1967.

10. An overview of the numerous findings can be found, for example, in S. Droit-Volet and S. Gil, "The Time-Emotion Paradox," *Philosophical Transactions of the Royal Society B* 364 (2009): 1943–1954.

11. A. Lambrechts, N. Mella, V. Pouthas, and M. Noulhiane, "Subjectivity of Time Perception: A Visual Emotional Orchestration," *Frontiers in Integrative Neuroscience* 5, no. 73 (2011); J. Wackermann, K. Meissner, D. Tankersley, and M. Wittmann, "Effects of Emotional Valence and Arousal on Acoustic Duration Reproduction Assessed via the 'Dual Klepsydra Model,'" *Frontiers in Neurorobotics* 8, no. 11 (2014).

12. V. van Wassenhove, M. Wittmann, A. D. Craig, and M. P. Paulus, "Psychological and Neural Mechanisms of Subjective Time Dilation," *Frontiers in Neuroscience* 5, no. 56 (2011).

13. C. Stetson, M. P. Fiesta, and D. M. Eagleman, "Does Time Really Slow Down during a Frightening Event?" *PLoS One* 2 (2007): e1295.

14. R. Buckley, "Slow Time Perception Can Be Learned," *Frontiers in Psychology* 5, no. 209 (2014).

15. D. Zakay and R. A. Block, "Temporal Cognition," *Current Directions in Psychological Science* 6 (1997): 12–16; D. Avni-Babad and I. Ritov, "Routine and the Perception of Time," *Journal of Experimental Psychology: General* 132 (2003): 543–550; M. Wittmann and S. Lehnhoff, "Age Effects in Perception of Time," *Psychological Reports* 97 (2005): 921–935.

16. Both the momentarily felt time expansion and retrospective memory are relevant factors in this discussion, as empirical research has shown. For example, one study initially showed that emotional stimuli are experienced as relatively longer lasting. This finding correlates with those of many other studies. Emotional stimuli could also be remembered better in retrospect. Prospectively, living in the moment accordingly results in time expansion, and retrospectively emotional stimuli are remembered better than more neutral stimuli. G. Dirnberger, G. Hesselmann, J. P. Roiser, S. Preminger, M. Jahanshahi, and R. Paz, "Give It Time: Neural Evidence for Distorted Time Perception and Enhanced Memory Encoding in Emotional Situations," *NeuroImage* 63 (2012): 591–599.

17. L. A. Campbell and R. A. Bryant, "How Time Flies: A Study of Novice Skydivers," *Behavior Research and Therapy* 45 (2007): 1389–1392.

18. O. Pollatos, J. Laubrock, and M. Wittmann, "Interoceptive Focus Shapes the Experience of Time," *PLoS One* 9 (2014): e86934.

19. Walter Benjamin, *On Hashish*, trans. Howard Eiland and others (Cambridge, MA, and London: Belknap Press of Harvard University Press, 2006), 49.

20. Cf. Wilhelm Schmid, *Mit sich selbst befreundet sein*, 6th ed. (Frankfurt and Main: Suhrkamp, 2013), 293.

21. N. Solowij and R. Battisti, "The Chronic Effects of Cannabis on Memory in Humans: A Review," *Current Drug Abuse Reviews* 1 (2008): 81–98.

22. R. A. Sewell, A. Schnakenberg, J. Elander, R. Radhakrishnan, A. Williams, P. D. Skosnik, B. Pittman, M. Ranganathan, and D. C. D'Souza, "Acute Effects of THC on Time Perception in Frequent and Infrequent Cannabis Users," *Psychopharmacology* 226 (2013): 401–413.

23. R. Ogden and C. Montgomery, "High Time," *Psychologist* 25 (2012): 590–592.

24. Charles Baudelaire, "The Poison," in *Flowers of Evil and Ennui*, trans. John E. Tidball (Bristol: Bishopston Editions, 2017), 137.

25. Ernest Jones, *The Life and Work of Sigmund Freud*, vol. 1, *(1856–1900)* (New York: Basic Books, 1953), 84. Freud's euphoria due to the effect of cocaine, for example as an anesthetic, evaporated a few years later when its addictive potential was recognized.

26. M. T. Fillmore, T. H. Kelly, and C. A. Martin, "Effects of D-amphetamine in Human Models of Information Processing and Inhibitory Control," *Drug and Alcohol Dependence* 77 (2005): 151–159.

27. M. Wittmann, *Felt Time* (Cambridge, MA: MIT Press, 2016).

28. Clearly written introductions to the way that, in physics, time has "disappeared" or has been "spatialized" can be found in L. Smolin, *Time Reborn* (London: Penguin Books, 2013), and in D. Buonomano, *Your Brain Is a Time Machine* (New York: W. W. Norton, 2017).

29. M. Bonato, M. Zorzi, and C. Umiltà, "When Time Is Space: Evidence for a Mental Time Line," *Neuroscience and Biobehavioral Reviews* 36 (2012): 2257–2273; R. Ulrich, V. Eikmeier, I. de la Vega, S. R. Fernández, S. Alex-Ruf, and C. Maienborn, "With the Past Behind and the Future Ahead: Back-to-Front Representation of Past and Future Sentences," *Memory & Cognition* 40 (2012): 483–495.

30. As to how language and culture shape the sense of time, see: L. Boroditsky, "How Language Shapes Thought," *Scientific American* 304 (2011): 62–65. In the language of the Aymara people of South America, the direction of time is reversed. As gestures and language show, for the Aymara the past lies ahead of them and the future behind them. It's actually quite logical: what is past I know, what is in the future, I don't know, so therefore I can't see it either, just like the things hiding behind my back. A good article on this subject can be found in *Science Daily*: "Backs to the Future: Aymara Language and Gesture Point to Mirror-Image View of Time," June 13, 2006. https://www.sciencedaily.com/releases/2006/06/060613185239.htm. Accessed December 27, 2017.

31. K. Stocker, "The Theory of Cognitive Spacetime," *Metaphor and Symbol* 29 (2014): 71–93; B. Kyu Kim, G. Zauberman, and J. R. Bettman, "Space, Time, and Intertemporal Preferences," *Journal of Consumer Research* 39 (2012): 867–880.

32. This retrospective aspect is discussed by Immanuel Kant in his *Anthropology from a Pragmatic Point of View*, what we might call his psychology textbook. In it he explains the phenomenon through the retrospective, memory-induced viewpoint: "The abundance of objects seen (villages and farmhouses) produces in our memory the deceptive conclusion that a vast amount of space has been covered and, consequently, that a longer period of time necessary for this purpose has also passed. However, the emptiness in the latter case produces little recollection of what has been seen and therefore leads to the conclusion that the route was shorter, and hence the time less, than would be shown by the clock." I. Kant, *Anthropology from a Pragmatic Point of View*, trans. and ed. Robert B. Loudon (Cambridge: Cambridge University Press, 2006), 130.

33. A. Saj, O. Fuhrman, P. Vuilleumier, and L. Boroditsky, "Patients with Left Spatial Neglect Also Neglect the 'Left Side' of Time," *Psychological Science* 25 (2013): 207–214.

34. See Henri Bergson's "Essai sur les données immédiates de la conscience," originally published in 1889: *Time and Free Will: An Essay on the Immediate Data of Consciousness*, trans. F. L. Pogson (London: George Allen & Co., 1913; repr. Dover, 2001), in particular pp. 104, 106, 107, 137, 224.

35. V. van Wassenhove, "Minding Time in an Amodal Representational Space," *Philosophical Transactions of the Royal Society B* 364 (2009): 1815–1830.

36. Thomas De Quincey, *Confessions of an English Opium-Eater* (London: Penguin, 2003), 76.

37. M. Wittmann and M. P. Paulus, "Decision Making, Impulsivity, and Time Perception," *Trends in Cognitive Sciences* 12 (2008): 7–12; M. Wittmann and M. P. Paulus, "How the Experience of Time Shapes Decision-making," in M. Reuter and C. Montag (eds.), *Neuroeconomics*. Studies in Neuroscience, Psychology and Behavioral Economics. (Berlin and Heidelberg: Springer, 2016): 133–144.

38. M. A. Sayette, G. Loewenstein, T. R. Kirchner, and T. Travis, "Effects of Smoking Urge on Temporal Cognition," *Psychology of Addictive Behaviors* 19 (2005): 88–93.

39. M. Wittmann, D. Leland, J. Churan, and M. P. Paulus, "Impaired Time Perception and Motor Timing in Stimulant-Dependent Subjects," *Drug and Alcohol Dependence* 90 (2007): 183–192.

40. J. Monterosso and G. Ainslie, "The Behavioral Economics of Will in Recovery from Addiction," *Drug and Alcohol Dependence* 90 (2007): S100–S111; K. N. Kirby, N. M. Petry, and W. K. Bickel, "Heroin Addicts Have Higher Discount Rates for Delayed Rewards Than Non-drug-using Controls," *Journal of Experimental Psychology: General* 128 (1999): 78–87.

41. D. Vaitl, N. Birbaumer, J. Gruzelier, G. A. Jamieson, B. Kotchoubey, A. Kübler, D. Lehmann, W. H. R. Miltner, U. Ott, P. Pütz, G. Sammer, I.

Strauch, U. Strehl, J. Wackermann, and T. Weiss, "Psychobiology of Altered States of Consciousness," *Psychological Bulletin* 131 (2005): 98–127.

42. P. Marshall, *Mystical Encounters with the Natural World* (Oxford: Oxford University Press, 2005); U. Ott, "Time Experience during Mystical States," in A. Nikolaidis and W. Achtner (eds.), *The Evolution of Time: Studies of Time in Science, Anthropology, Theology* (Oak Park, IL: Bentham Science Publishers, 2013), 104–116; B. Shanon, "Altered Temporality," *Journal of Consciousness Studies* 8 (2001): 35–58.

43. Oliver Davies (trans. and ed.), *Meister Eckhart: Selected Writings* (London: Penguin, 1994), 147.

44. E. Tugendhat, *Egocentricity and Mysticism: An Anthropological Study* (New York: Columbia University Press, 2016), 95.

45. One expression of this egocentricity is surely the notorious ideas of religiously motivated suicide bombers in the Middle East, who are promised palaces and virgins after death. Another story testifies to the same kind of egocentricity. A man wanted to kill himself by throwing himself off the Golden Gate Bridge in San Francisco. When he hit the water, he broke both knees, but survived. He thrashed about helplessly on the surface of the water, threatening to go under. His survival was ensured by a seal, which swam up and kept prodding him above water with its nose until the rescue services arrived. What was his reaction to being rescued by a seal? Afterward, no, he didn't take up the cause of rescuing and looking after seals at California's rescue centers for distressed seals, but instead he became intensely religious. Ultimately, God had saved *him*; out of all the billions of people, God had sent *him* a sign.

46. T. Strässle, *Gelassenheit. Über eine andere Haltung zur Welt* (Munich: Hanser, Edition Akzente, 2013), 36ff. In the words of the British philosopher John Gray: "Spiritual life is not a search for meaning but a release from it." J. Gray, *Stray Dogs* (London: Granta Books, 2002).

47. For a discussion of the ideas of the theologian Rudolf Otto and the psychologist William James, see D. M. Wulf, "Mystical Experiences," in

E. Cardena, S. J. Lynn, and S. Krippner (eds.), *Varieties of Anomalous Experience: Examining the Scientific Evidence*, 2nd ed., (Washington, DC: American Psychological Association, 2014), 369–408. William James cites many vivid reports of spontaneous mystical experiences in his work *The Varieties of Religious Experience* (London: Penguin Classics, 1983). An overview of the common core in mystical experiences can also be found in A. Geels, "Altered Consciousness in Religion," in E. Cardena and M. Winkelman (eds.), *Altering Consciousness: Multidisciplinary Aspects*, vol. 1, *History, Culture, and the Humanities* (Santa Barbara, CA: Praeger, 2011), 205–276.

48. The concept of "the encompassing" can be found in Karl Jaspers, *Way to Wisdom: An Introduction to Philosophy*, trans. Ralph Manheim (New Haven, CT: Yale University Press, 2003), 33–34.

49. W. Achtner, "Time, Eternity, and Trinity," *Neue Zeitschrift für Systematische Theologie und Religionsphilosophie* 51 (2009): 267–288.

50. J. Kiverstein, "The Minimal Sense of Self, Temporality and the Brain," *Psyche* 15 (2009): 59–74; M. Wittmann, "Embodied Time: The Experience of Time, the Body, and the Self," in V. Arstila and D. Lloyd (eds.), *Subjective Time: The Philosophy, Psychology and Neuroscience of Temporality* (Cambridge, MA: MIT Press, 2014): 507–523. In the words of Martin Heidegger, "The ecstatic unity of temporality—that is, the unity of the 'outside-itself' in the raptures of the future, the having-been, and the present—is the condition of the possibility that there can be a being that exists as its 'There.' The being that bears the name Da-sein is 'cleared.'" M. Heidegger, *Being and Time*, trans. Joan Stambaugh (New York: State University of New York Press, 1965), 321.

51. "There was no past and no future, only awareness of living in an eternal moment that encompassed all that has been, that is, and that will be." These are the words of an otherwise unspiritual person describing their altered experience of life after a sudden mystical experience. Quoted in P. Marshall, *Mystical Encounters with the Natural World* (Oxford: Oxford University Press, 2005), 25. A longer essay on the opportunities for encouraging extraordinary states of consciousness in life and leading a more intense and fulfilled life in the long term is

Colin Wilson's book *Super Consciousness: The Quest for the Peak Experience* (London: Watkins Publishing, 2009).

52. T. Schäfer, J. Fachner, and M. Smukalla, "Changes in the Representation of Space and Time while Listening to Music," *Frontiers in Psychology* 4 (2013): 508.

53. Egon Voss, *Richard Wagner: Tristan und Isolde* (Stuttgart: Reclam, 2003), 119.

54. As Martin Heidegger wrote, following an idea of Wilhelm Dilthey: "What is real is experienced in impulse and will. Reality is resistance, more exactly the character of resistance." M. Heidegger, *Being and Time*, trans. Joan Stambaugh (New York: State University of New York Press, 1965), 194.

55. Arthur Schopenhauer's pessimistic worldview can be summed up as a pendulum swinging backward and forward between pain and boredom. The poor individual starts out hungry and cold; if they ultimately attain riches, at best they feel the bliss of excess. In Schopenhauer's view, wanting and desiring is the condition of lack. However, the desired condition brings boredom.

56. W. Achtner, "Time, Eternity, and Trinity," *Neue Zeitschrift für Systematische Theologie und Religionsphilosophie* 51 (2009): 267–288.

57. Carl Albrecht, *Psychologie des mystischen Bewusstseins* (Bremen: Carl Schünemann, 1951), 78.

58. B. Shanon, "Altered Temporality," *Journal of Consciousness Studies* 8 (2001): 35–58.

59. Ibid.; Z. Mishor, D. J. McKenna, and J. C. Callaway, "DMT and Human Consciousness," in E. Cardena and M. Winkelman (eds.), *Altering Consciousness: Multidisciplinary Aspects*, vol. 2, *Biological and Psychological Perspectives* (Santa Barbara, CA: Praeger, 2011), 85–119.

60. Aldous Huxley, *The Doors of Perception* (New York: Harper Collins, 2009), 21.

61. P. Nádas, *Own Death*, trans. János Salomon (Göttingen: Steidl, 2006).

62. P. van Lommel, *Consciousness beyond Life: the Science of the Near-Death Experience*, trans. Laura Vroomen (New York: Harper Collins, 2011).

63. A recently published multicenter clinical study systematically collected near-death experiences of people after heart attacks: S. Parnia et al., AWARE—AWAreness during REsuscitation—A Prospective Study," *Resuscitation* 85 (2014), 1799–1805. In a pioneering study assessing the experience of time during near death, 74 percent of respondents replied with the prespecified item "Everything seemed to be happening all at once; or time stopped, or lost all meaning": B. Greyson, "Near-Death Encounters with and without Near-Death Experiences: Comparative NDE Scale Profiles," *Journal of Near-Death Studies* 8 (1990): 151–161. We just published a further analysis of reports from a publicly available web-based data bank of the *Near Death Experience Research Foundation* and found that 65 percent of individuals who had a NDE report of a "change of subjective time." Of those people experiencing a change in subjective time, 94% report the feeling of "timelessness": M. Wittmann, L. Neumaier, R. Evrard, A. Weibel, I. Schmied-Knittel, "Subjective Time Distortion during Near-Death Experiences: An Analysis of Reports." *Zeitschrift für Anomalistik* 17 (2017): 309–320.

64. H. Knoblauch, I. Schmidt-Knittel, and B. Schnettler, "The Different Experience: A Report on a Survey of Near-Death Experiences in Germany," *Journal of Near-Death Studies* 20 (2001): 15–29.

65. Reports of such experiences are known from pre-Christian history, ranging from Sumerian to Roman and classical Greek culture: M. Nahm, "Four Ostensible Near-Death Experiences of Roman Times with Peculiar Features: Mistake Cases, Correction Cases, Xenoglossy, and a Prediction," *Journal of Near-Death Studies* 27 (2009): 211–222. They are also mentioned in the Bible (2 Corinthians 12:1–4). My colleague Lukasz Smigielski of Zurich University has drawn my attention to the fact that both *The Tibetan Book of the Dead* and Plato's *Republic* (*Politeia*) include descriptions of near-death experiences.

66. A summary of the possibilities of interpretation in neurophysiology, adopting a priori the materialistic position with almost evangelical

fervor, is D. Mobbs and C. Watt, "There Is Nothing Paranormal about Near-Death Experiences: How Neuroscience Can Explain Seeing Bright Lights, Meeting the Dead, or Being Convinced You Are One of Them," *Trends in Cognitive Sciences* 15 (2010): 447–449.

67. This is the argument put forward by Pim van Lommel in his book *Consciousness beyond Life* (New York: Harper Collins, 2011).

68. Typing the keywords "Pam Reynolds" and "near-death experience" into an Internet search engine yields many blog posts and other sources. The cardiologist Michael Sabom has researched this and other cases and written a book: M. Sabom, *Light and Death: One Doctor's Fascinating Account of Near-Death Experiences* (Grand Rapids: Zondervan, 1998). A short scientific treatment of the case can be found in M. Beauregard, "Transcendent Experiences and Brain Mechanisms," in E. Cardena and M. Winkelman (eds.), *Altering Consciousness: Multidisciplinary Aspects*, vol. 2, *Biological and Psychological Perspectives* (Santa Barbara, CA: Praeger, 2012), 63–84.

69. A scientific debate about the Reynolds case can be found in the *Journal of Near-Death Studies* 25, no. 4 (2007).

Chapter 2

1. Karl Jaspers, *Psychologie der Weltanschauungen*, 3rd ed. (Berlin: Julius Springer, 1925), 112.

2. See, for example, the ideas of the sociologist Hartmut Rosa in *Social Acceleration: A New Theory of Modernity*, trans. Jonathan Trejo-Mathys (New York: Columbia University Press, 2013). For the psychological and neuroscientific view on this, see M. Wittmann, *Felt Time*, trans. Erik Butler (Cambridge, MA: MIT Press, 2016).

3. Byung-Chul Han, *The Scent of Time: A Philosophical Essay on the Art of Lingering*, trans. Daniel Steuer (Cambridge: Polity, 2017), 37–38.

4. The French philosopher of Romanian origin E. M. Cioran, who is almost unsurpassed in his pessimism, had his own ecstatic, mystical experiences without descending into transcendental effusiveness as a

result. He too describes the two qualities of duration and moment: "The experience of eternity therefore depends on intensity of subjective feeling, and the way to eternity is to transcend the temporal. One must fight hard against time so that—once the mirage of the succession of moments is overcome—one can live fully the instant one is launched into eternity. ... those with a keen consciousness of temporality live every moment thinking of the next one. Eternity can be attained only if there are no connections, if one lives the instant totally and absolutely." E. M. Cioran, *On the Heights of Despair*, trans. with an introduction by Ilinca Zarifopol-Johnston (Chicago and London: University of Chicago Press, 1992), 64.

5. Ludwig Wittgenstein, *Tractatus Logico-Philosophicus* (Oxford: Routledge, 2001), 87.

6. Epicurus (341–270 BCE) wrote a letter to Menoeceus in which he wants to convince his correspondent that "death is of no concern to us." Epicurus, *Letters, Principle Doctrines and Vatican Sayings*, trans. with an introduction and notes by Russel M. Geer (Indianapolis and New York: Bobbs-Merrill Co., 1964), 54.

7. M. Wittmann and M. Paulus, "How the Experience of Time Shapes Decision-Making," in M. Reuter and C. Montag (eds.), *Neuroeconomics. Studies in Neuroscience, Psychology and Behavioral Economics.* (Berlin, Heidelberg: Springer, 2016): 133–144.

8. These ideas can be found in the work of Karl Jaspers, Martin Heidegger, Søren Kierkegaard, Jean-Paul Sartre, and Otto Friedrich Bollnow.

9. Peter Handke, *To Duration*, trans. Scott Abbott (Amsterdam: The Last Books, 2015), 13.

10. Saint Augustine discussed the question of the possible duration of an instant in such a substantial way that no philosopher has been able to ignore his ideas ever since. For an account of his ideas see K. Flasch, *Was ist Zeit? Augustinus von Hippo. Das XI . Buch der Confessiones* (Frankfurt and Main: Vittorio Klostermann, 1993). In the nineteenth century, the British empiricists wrestled with and further developed the notion of the moment of time, until in 1890 William James formulated the idea

of the "specious present" as an expanded moment. This historical development is described in H. Andersen, "The Development of the 'Specious Present' and James's Views on Temporal Experience," in V. Arstila and D. Lloyd (eds.), *Subjective Time: The Philosophy, Psychology, and Neuroscience of Temporality* (Cambridge, MA: MIT Press, 2014), 25–42.

11. Moreover, this shows how the representation of temporal relationships is dependent on the spatial organization of moments in time. Time is conceptualized as a spatial arrow of time. Furthermore, one could also analyze linguistically the various expressions for present experience—the instant, the moment, or presence—and ask whether and how they are the same and in what ways they are different, also in reference to the question of temporal duration.

12. On the difference between the physical and the experiential concept of time, see C. Callender, *What Makes Time Special?* (Oxford: Oxford University Press, 2017); N. D. Mermin, "QBism Puts the Scientist Back into Science," *Nature* 507 (2014): 421–423.

13. The works of Edmund Husserl, Maurice Merleau-Ponty, and William James revolve around this dual aspect of time consciousness. See M. Wittmann, "The Subjective Flow of Time," in H. J. Birks (ed.), *The Encyclopedia of Time* (Thousand Oaks, CA: SAGE Publications, 2009), 1322–1324.

14. In reference to the narrative self, see S. Gallagher, "Philosophical Conceptions of the Self: Implications for Cognitive Science," *Trends in Cognitive Science* 4 (2000): 14–21.

15. I examine the various temporal levels of the duration of the moment in more depth in M. Wittmann, *Felt Time* (Cambridge, MA: MIT Press, 2016). A scientific treatment of this can be found in M. Wittmann, "Moments in Time," *Frontiers in Integrative Neuroscience* 5, no. 66 (2011).

16. E. Husserl, *On the Phenomenology of the Consciousness of Internal Time (1893–1917)*, trans. John Barnett Brough (Dordrecht, Boston, and London: Kluwer Academic Publishers, 1991).

17. K. Friston, "The Free-Energy Principle: A Unified Brain Theory?" *Nature Reviews Neuroscience* 11 (2010): 127–138. The predictive coding theory thereby is based on the earlier idea of the reafference principle: E. von Holst and H. Mittelstaedt, "Das Reafferenzprinzip," *Naturwissenschaften* 37 (1950): 464–476. In extreme situations of exertion and dedication, protention can almost seem like a paranormal glimpse of the future. For example, professional footballers describe what is for them the unique event of their career, when they anticipated the moves before the goal was scored and the event happened in exactly that way a short time later. See Zinédine Zidane's brief commentary in the movie *Zidane* when, during a match with his club Real Madrid, the footballer was filmed by 17 cameras, which showed Zidane alone during the entire game. Another example: The Algerian footballer Salah Assad, who was on the pitch during his national team's victory over Germany in 1982, also had this extraordinary experience, as he described in an interview with the *Süddeutsche Zeitung* (June 30, 2014): "SZ: What is your most powerful memory of that day? Assad: I still have a clear memory of the second goal. Because I already had this goal in my head, before it was scored. I saw it, and it happened in exactly the same way. That was extraordinary."

18. S. Gallagher and D. Zahavi, "Primal Impression and Enactive Perception," in V. Arstila and D. Lloyd (eds.), *Subjective Time* (Cambridge, MA: MIT Press, 2014), 83–99.

19. J. Kiverstein, "Making Sense of Phenomenal Unity: An Intentionalist Account of Temporal Experience," *Royal Institute of Philosophy Supplement* 85 (2010): 155–181. The philosopher Dan Lloyd has wrapped up an explanation of Husserl's ideas in an exciting detective novel: D. Lloyd, *Radiant Cool: A Novel Theory of Consciousness* (Cambridge, MA: MIT Press, 2004). Alongside this intentional theory of temporal experience in which the co-present orientation of experience toward what is just past, present, and future creates time consciousness, there is also the extensional theory of temporal experience, which assumes that consciousness expands in time and that this temporal expansion corresponds to the temporal sequence of world events. Here, this extension of present-time consciousness has a specific duration. See also this essay:

C. Hoerl, "Time and Tense in Perceptual Experience," *Philosopher's Imprint* 9, no. 12 (2009); and the book by B. Dainton, *Time and Space* (Chesham: Acumen, 2001). For a discussion of the approaches, see J. Kiverstein and V. Arstila, "Time in Mind," in H. Dyke and A. Bardon (eds.), *A Companion to the Philosophy of Time* (Oxford: John Wiley & Sons, 2013). An overview of the various theories can be found in B. Dainton, "Temporal Consciousness" (2010; rev. 2017), in the online *Stanford Encyclopedia of Philosophy*, ed. E. N. Zalta. http://plato.stanford .edu/entries/consciousness-temporal/. Accessed December 28, 2017.

20. E. Pöppel, *Mindworks: Time and Conscious Experience*, trans. Tom Artin (New York: Harcourt Brace Jovanovich Inc., 1988); E. Pöppel, "Pre-semantically Defined Temporal Windows for Cognitive Processing," *Philosophical Transactions of the Royal Society B* 364 (2009): 1887–1896; E. Pöppel and Y. Bao, "Temporal Windows As a Bridge from Objective to Subjective Time," in V. Arstila and D. Lloyd (eds.), *Subjective Time* (Cambridge, MA: MIT Press, 2014), 241–262. A conceptual debate is currently under way, with regard to empirical findings, to decide whether the experienced moment or the feeling of "nowness" stems from one temporal processing mechanism with a defined duration of around 2 to 3 seconds or whether several independent mechanisms with variable duration—but nevertheless in the range of seconds—can be identified: P. A. White, "The Three-Second "Subjective Present": A Critical Review and a New Proposal," *Psychological Bulletin* 143 (2017): 735–756.

21. M. Wittmann and E. Pöppel, "Temporal Mechanisms of the Brain as Fundamentals of Communication—With Special Reference to Music Perception and Performance," *Musicae Scientiae*, special issue 1999–2000 (1999): 13–28; W. Tschacher, F. Ramseyer, and C. Bergomi, "The Subjective Present and Its Modulation in Clinical Contexts," *Timing & Time Perception* 1 (2013): 239–259.

22. M. Fink, P. Ulbrich, J. Churan, and M. Wittmann, "Stimulus-Dependent Processing of Temporal Order," *Behavioral Processes* 71 (2006): 344–352; I. J. Hirsh, and C. Sherrick, "Perceived Order Indifferent Sense Modalities," *Journal of Experimental Psychology* 62 (1961): 423–432.

23. P. Ulbrich, J. Churan, M. Fink, and M. Wittmann, "Perception of Temporal Order: The Effects of Age, Sex, and Cognitive Factors," *Aging, Neuropsychology and Cognition* 16 (2009): 183–202; R. M. Warren and C. J. Obusek, "Identification of Temporal Order within Auditory Sequences," *Perception & Psychophysics* 12 (1972): 86–90.

24. D. Avni-Babad and I. Ritov, "Routine and the Perception of Time," *Journal of Experimental Psychology: General* 132 (2003): 543–550; M. Wittmann and S. Lehnhoff, "Age Effects in the Perception of Time," *Psychological Reports* 97 (2005): 921–935.

25. See the chapter "The 'Sense' of Time Passing," in J. Kabat-Zinn, *Coming to Our Senses: Healing Ourselves and the World through Mindfulness* (New York: Hyperion, 2006), 162–166, here 163.

26. For this historical and thematic derivation I have drawn on conversations with Stefan Schmidt of the Freiburg University Medical Center. The source of this information is V. Analayo, *Satipatthana: The Direct Path to Realization* (Cambridge: Windhorse, 2004).

27. J. Kabat-Zinn, *Wherever You Go, There You Are: Mindfulness Meditation in Everyday Life* (New York: Hyperion, 1994).

28. S. Schmidt, "Opening up Meditation for Science: The Development of a Meditation Classification System," in S. Schmidt and H. Walach (eds.), *Meditation: Neuroscientific Approaches and Philosophical Implications*. Studies in Neuroscience, Consciousness and Spirituality 2 (Cham, Switzerland: Springer, 2014), 137–152.

29. For a broader consideration of these ideas see M. Wittmann and S. Schmidt, "Mindfulness Meditation and the Experience of Time," in S. Schmidt and H. Walach (eds.), *Meditation: Neuroscientific Approaches and Philosophical Implications*. Studies in Neuroscience, Consciousness and Spirituality 2 (Cham, Switzerland: Springer, 2014), 199–210.

30. A. Lutz, H. A. Slagter, J. D. Dunne, and R. J. Davidson, "Attention Regulation and Monitoring in Meditation," *Trends in Cognitive Science* 12 (2008): 163–169.

31. K. A. MacLean, E. Ferrer, S. R. Aichele, D. A. Bridwell, A. P. Zanesco, T. L. Jacobs, B. G. King, E. L. Rosenberg, B. K. Sahdra, P. R. Shaver, B. A. Wallace, G. R. Mangun, and C. D. Saron, "Intensive Meditation Training Improves Perceptual Discrimination and Sustained Attention," *Psychological Science* 21 (2010): 829–839.

32. A. P. Jha, J. Krompinger, and M. J. Baime, "Mindfulness Training Modifies Subsystems of Attention," *Cognitive, Affective, and Behavioral Neuroscience* 7 (2007): 109–119.

33. F. Zeidan, S. K. Johnson, B. J. Diamond, Z. David, and P. Goolkasian, "Mindfulness Training Improves Cognition: Evidence of Brief Mental Training," *Consciousness and Cognition* 19 (2010): 597–605.

34. M. D. Mrazek, M. S. Franklin, D. T. Phillips, B. Baird, and J. W. Schooler, "Mindfulness Training Improves Working Memory Capacity and GRE Performance While Reducing Mind Wandering," *Psychological Science* 24 (2013): 776–781.

35. This was a collaborative project by three research groups led by Harald Walach and Ursula Mochty of Viadrina European University in Frankfurt an der Oder, Sebastian Sauer and Niko Kohls at the Generation Research Program, Bad Tölz, of the Ludwig-Maximilian University in Munich, and Marc Wittmann at the Institute for Frontier Areas of Psychology and Mental Health in Freiburg. The project was funded by the BIAL Foundation in Porto under the title "A Test for Mindfulness: The Bistable Images Test," research funding awarded to Prof. Harald Walach and Dr. Ursula Mochty. The results of the study can be read in S. Sauer, J. Lemke, M. Wittmann, N. Kohls, U. Mochty, and H. Walach, "How Long is Now for Mindfulness Meditators?" *Personality and Individual Differences* 52 (2012): 750–754.

36. A doctoral thesis is available online in which Jannis Wernery of the Collegium Helveticum in Zurich assessed mindfulness as an individual personality variable and found correlations between the reversal intervals when looking at a Necker cube and mindfulness. The more mindful the student test subjects were, the slower the reversal perceived in the Necker cube. J. Wernery, *Bistable Perception of the Necker Cube in the Con-*

text of Cognition and Personality, doctoral thesis at the ETH Zurich (2013): http://dx.doi.org/10.3929/ethz-a-009900582. A more recent study complements these findings: J. Kornmeier, E. Friedel, M. Wittmann, and H. Atmanspacher, "EEG Correlates of Cognitive Time Scales in the Necker-Zeno Model for Bistable Perception," *Consciousness and Cognition* 53 (2017): 136–150.

37. R. S. S. Kramer, U. W. Weger, and D. Sharma, "The Effect of Mindfulness Meditation on Time Perception," *Consciousness and Cognition* 22 (2013): 846–852. Since then work conducted in particular by the French psychologist Sylvie Droit-Volet has confirmed the results of an expansion of time in meditation-naive and meditation-experienced individuals: S. Droit-Volet, M. Fanget, and M. Dambrun, "Mindfulness Meditation and Relaxation Training Increases Time Sensitivity," *Consciousness and Cognition* 31 (2015): 86–97.

38. These studies were conducted jointly by Marc Wittmann at the Institute for Frontier Areas of Psychology and Mental Health in Freiburg and by Karin Meissner at the Institute of Medical Psychology at the Ludwig-Maximilian University in Munich, and have been published in M. Wittmann, J. Peter, O. Gutina, S. Otten, N. Kohls, and K. Meissner, "Individual Differences in Self-attributed Mindfulness Levels are Related to the Experience of Time and Cognitive Self-control," *Personality and Individual Differences* 64 (2014): 41–45.

39. An overview of the current tools for self-assessing mindfulness can be found in S. Sauer, H. Walach, S. Schmidt, T. Hinterberger, S. Lynch, A. Büssing, and N. Kohls, "Assessment of Mindfulness: Review on State of the Art," *Mindfulness* 4 (2013): 3–17.

40. H. Walach, N. Buchheld, V. Buttenmüller, N. Kleinknecht, and S. Schmidt, "Measuring Mindfulness: The Freiburg Mindfulness Inventory (FMI)," *Personality and Individual Differences* 40 (2006): 1543–1555.

41. The *Comprehensive Inventory of Mindfulness Experiences* (CHIME): C. Bergomi, W. Tschacher, and Z. Kupper, "Measuring Mindfulness: First Steps towards the Development of a Comprehensive Mindfulness Scale," *Mindfulness* 4 (2013): 18–32.

42. The German versions of the Barratt Impulsiveness Scale and the Zimbardo Time Perspective Inventory were used.

43. Equally, other studies show how more mindful people also appear less impulsive in objective behavioral tests: S. Sauer, H. Walach, and N. Kohls, "Gray's Behavioural Inhibition System as a Mediator of Mindfulness towards Well-being," *Personality and Individual Differences* 50 (2011): 506–511; Y-C. Lee and H-F. Chao, "The Role of Active Inhibitory Control in Psychological Well-being and Mindfulness," *Personality and Individual Differences* 53 (2012): 618–621.

44. T. Esch, "The Neurobiology of Meditation and Mindfulness," in S. Schmidt and H. Walach (eds.), *Meditation: Neuroscientific Approaches and Philosophical Implications*. Studies in Neuroscience, Consciousness and Spirituality 2 (Cham, Switzerland: Springer, 2014), 153–173; A. P. Jha, E. A. Stanley, A. Kiyonaga, L. Wong L, and L. Gelfand, "Examining the Protective Effects of Mindfulness Training on Working Memory Capacity and Affective Experience," *Emotion* 10 (2010): 54–64.

45. The influence of these three components on time perception is examined in detail in M. Wittmann and S. Schmidt, "Mindfulness Meditation and the Experience of Time," in S. Schmidt and H. Walach (eds.), *Meditation: Neuroscientific Approaches and Philosophical Implications*. Studies in Neuroscience, Consciousness and Spirituality 2 (Cham, Switzerland: Springer, 2014), 199–210.

46. Reviews of the three factors in time perception can be found in M. Wittmann, "The Inner Sense of Time: How the Brain Creates a Representation of Duration," *Nature Reviews Neuroscience* 14 (2013): 217–223; M. Wittmann, "The Inner Experience of Time," *Philosophical Transactions of the Royal Society B* 364 (2009): 1955–1967.

47. M. Sucala and D. David, "Mindful about Time in a Fast Forward World: The Effects of Mindfulness Exercise on Time Perception," *Transylvanian Journal of Psychology* 14 (2013): 243–253.

48. This study was funded by the BIAL Foundation in Porto: M. Wittmann, K. Meissner, and S. Schmidt, "The Embodied Experience of Time: Modulations of Mindfulness Meditation," Fundação BIAL, grant no.

52/12. See the related publication: M. Wittmann, S. Otten, E. Schötz, A. Sarikaya, H. Lehnen, H-G. Jo, N. Kohls, S. Schmidt, and K. Meissner, "Subjective Expansion of Extended Time-Spans in Experienced Meditators," *Frontiers in Psychology* 5 (2015): 1586.

49. See the classic book on an embodied neurophenomenology: F. J. Varela, E. Thompson, and E. Rosch, *The Embodied Mind: Cognitive Science and Human Experience* (Cambridge, MA: MIT Press, 1991; repr. 2017). For example, the research group set up by Stefan Schmidt of the University Clinic in Freiburg with Han-Gue Jo, Marc Wittmann, and Thilo Hinterberger, who are researching the topic of voluntary motor responses with the meditation master Tilmann Lhündrup Borghardt (TLB) using the famous Libet task. The studies show how measured neuronal activity before and during a motor action correlates with TLB's introspection. For example, the felt duration of an intention preceding a finger movement, corresponded with the underlying brain activity measured in EEG data: H.-G. Jo, M. Wittmann, T. L. Borghardt, T. Hinterberger, and S. Schmidt, "First-Person Approaches in Neuroscience of Consciousness: Brain Dynamics Correlate with the Intention to Act," *Consciousness and Cognition* 26 (2014): 105–116.

50. A. Berkovich-Ohana, Y. Dor-Ziderman, J. Glicksohn, and A. Goldstein, "Alterations in the Sense of Time, Space, and Body in the Mindfulness-Trained Brain: A Neurophenomenologically-guided MEG Study," *Frontiers in Psychology* 4 (2013): 912.

51. In March 2014 Stefan Schmidt, who is researching mindfulness meditation at the University of Freiburg, and I met with the doctor and meditation teacher Tilmann Lhündrup Borghardt at the Institute for Frontier Areas of Psychology and Mental Health.

52. As a Buddhist meditation teacher, the medical doctor Tilmann Borghardt uses the name Lama Tilmann Lhündrup.

53. T. Metzinger, *Being No One: The Self-Model Theory of Subjectivity* (Cambridge, MA: MIT Press, 2004). A popular-science version can be found in T. Metzinger, *The Ego Tunnel: The Science of the Mind and the Myth of the Self* (New York: Basic Books, 2009).

54. V. Dufour, C. A. F. Wascher, A. Braun, R. Miller, and T. Bugnyar, "Corvids Can Decide If a Future Exchange Is Worth Waiting For," *Biology Letters* 8 (2012): 201–204; J. M. Allman, N. A. Tetreault, A. Y. Hakeem, K. F. Manaye, K. Semendeferi, J. M. Erwin, S. Park, V. Goubert, and P. R. Hof, "The von Economo Neurons in Frontoinsular and Anterior Cingulate Cortex in Great Apes and Humans," *Brain Structure and Function* 214 (2010): 495–517; A. D. Craig, "How Do You Feel—Now? The Anterior Insula and Human Awareness," *Nature Reviews Neuroscience* 10 (2009): 59–68.

55. J. Kiverstein, "Making Sense of Phenomenal Unity: An Intentionalist Account of Temporal Experience," *Royal Institute of Philosophy Supplement* 85 (2010): 155–181; D. Lloyd, *Radiant Cool: A Novel Theory of Consciousness* (Cambridge, MA: MIT Press, 2004). See also my essay, M. Wittmann, "Embodied Time: The Experience of Time, the Body, and the Self," in V. Arstila and D. Lloyd (eds.), *Subjective Time* (Cambridge, MA: MIT Press, 2014), 507–523.

56. See the phenomenological line of argument in S. Gallagher and D. Zahavi, "Primal Impression and Enactive Perception," in V. Arstila and D. Lloyd (eds.), *Subjective Time* (Cambridge, MA: MIT Press, 2014), 83–99.

57. M. Dambrun, "When the Dissolution of Perceived Body Boundaries Elicits Happiness: The Effect of Selflessness Induced by a Body Scan Meditation," *Consciousness and Cognition* 46 (2016): 89–98.

58. M. Dambrun and M. Ricard, "Self-Centeredness and Selflessness: A Theory of Self-Based Psychological Functioning and Its Consequences for Happiness," *Review of General Psychology* 15 (2011): 138–157.

59. The story of how the ring-shaped structure of the benzene molecule came to the chemist August Kekulé in a daydream is legendary. The mathematician Henri Poincaré described how he happened upon the idea that helped him make a breakthrough while getting on a bus, lost in thought. Similarly, product developers often have groundbreaking ideas anywhere but at their desks. Daydreaming and mind wandering seem to help in the synthesizing of ideas that emerge of their own accord, letting them float to the surface of consciousness—the eureka

moment. See Jonah Lehrer, *Imagine! How Creativity Works* (New York: Houghton Mifflin Harcourt, 2012), 46ff.

60. A review of this topic can be found in J. W. Schooler, M. D. Mrazek, M. S. Franklin, B. Baird, B. W. Mooneyham, C. Zedelius, and J. M. Broadway, "The Middle Way: Finding the Balance between Mindfulness and Mind-Wandering," in B. H. Ross (ed.), *The Psychology of Learning and Motivation*, vol. 60 (Burlington: Academic Press, 2014), 1–33. The title of this work makes it clear that both components—mindfulness and mind wandering—must work together in order for one to be focused on the one hand, and on the other to give the imagination free rein.

Chapter 3

1. G.-R. D'Allonnes, "Rôle des sensations internes dans les émotions et dans la perception de la durée," *Revue Philosophique de la France et de l'Étranger* 60 (1905): 592–623.

2. The strange case of Alexandrine is indeed a quite peculiar one, as d'Allonnes describes in detail the symptoms of the loss of physical, affective, and temporal perceptions—supported by experiments. We should highlight in particular the striking loss of the sense of time. The connection between a physical, emotional, and temporal perception disorder was, however, well known and reported, by the French psychiatrist Pierre Janet (1859–1947), for example. See also the descriptions of patients by the psychoanalyst Peter Hartocollis in his book *Time and Timelessness: The Varieties of Temporal Experience* (New York: International Universities Press, 1983), 111ff. Here Hartocollis classifies the reported disorders through a diagnosis of "depersonalization," which he lists under "psychotic disorders" in one of the chapters. Today, however, we distinguish depersonalization disorders from psychotic disorders; no delusions are present in the former.

3. Thanks to Johannes Angenvoort for drawing my attention to these lyrics from Chuck Berry's song *Little Queenie*.

4. See remarks by Richard Shusterman, who is researching the opposing notions of William James and Ludwig Wittgenstein with regard to the

conscious perception of self and body: R. Shusterman, *Body Consciousness: A Philosophy of Mindfulness and Somaesthetics* (Cambridge: Cambridge University Press, 2008).

5. Different variations of the "embodiment" approach are discussed in S. Gallagher, *Enactivist Interventions* (Oxford: Oxford University Press, 2017); one outstanding conceptual beginning of the embodiment "movement" is still F. J. Varela, E. Thompson, and E. Rosch, *The Embodied Mind: Cognitive Science and Human Experience* (Cambridge MA: MIT Press, 1991; repr. 2017).

6. However, it must be said that insights into the connection between time perception and perception of the body have been largely ignored only in the "Western world." In 1946, in his *Fundamentals of General Psychology* (original in Russian), in the chapter "The Perception of Time," the Soviet psychologist S. L. Rubinstein regarded "temporal sensitivity" to be "autonomic sensitivity" as a matter of course. The work also makes reference to Revault d'Allonnes, whose work I came across first via this source, that is, indirectly via English-language pages of a website, "Tempometry Lab," created by Russian colleagues: http://tempometry.ru/.

7. A. D. Craig, "Emotional Moments across Time: A Possible Neural Basis for Time Perception in the Anterior Insula," *Philosophical Transactions of the Royal Society B* 364 (2009): 1933–1942; A. D. Craig, "How Do You Feel—Now? The Anterior Insula and Human Awareness," *Nature Reviews Neuroscience* 10 (2009): 59–70.

8. The original article on this topic is M. Wittmann, A. N. Simmons, J. Aron, and M. P. Paulus, "Accumulation of Neural Activity in the Posterior Insula Encodes the Passage of Time," *Neuropsychologia* 48 (2010): 3110–3120. An overview can be found in M. Wittmann, "The Inner Sense of Time: How the Brain Creates a Representation of Duration," *Nature Reviews Neuroscience* 14 (2013): 217–223.

9. See issue no. 364 of the *Philosophical Transactions of the Royal Society B*, edited by Virginie von Wassenhove and me in 2009, in which leading neuroscientists present their models of temporal perception; as well as the introduction to the special issue: M. Wittmann and V. van Wassen-

hove, "Introduction. The Experience of Time: Neural Mechanisms and the Interplay of Emotion, Cognition and Embodiment," *Philosophical Transactions of the Royal Society B* 364 (2009): 1809–1813.

10. Maurice Merleau-Ponty, *Phenomenology of Perception*, trans. Colin Smith (London and New York: Routledge, 1962). See also the essay, V. Arstila and D. Lloyd, "Subjective Time: From Past to Future," in V. Arstila and D. Lloyd (eds.), *Subjective Time: The Philosophy, Psychology, and Neuroscience of Temporality* (Cambridge, MA: MIT Press, 2014), 309–321. The philosopher Yvonne Förster-Beuthan also discusses Merleau-Ponty's approach in the context of other philosophers of time: Y. Förster-Beuthan, *Zeiterfahrung und Ontologie. Perspektiven moderner Zeitphilosophie* (Munich: Wilhelm Fink, 2012).

11. J. D. Watt, "Effect of Boredom Proneness on Time Perception," *Psychological Reports* 69 (1991): 323–327.

12. R. Safranski, *Martin Heidegger: Between Good and Evil*, trans. Ewald Osers (Cambridge, MA: Harvard University Press, 1998), 192, 194.

13. This is the Freiburg lecture series of the winter semester 1929/30: M. Heidegger, *The Fundamental Concepts of Metaphysics: World, Finitude, Solitude*, trans. William McNeill and Nicholas Walker (Bloomington, IN: Indiana University Press, 1995), 80.

14. Ibid., 143. On pp. 163–64, Heidegger relates "boredom" to the loss of "mystery," as it were, the loss of the meaning of life as a sense of the all-encompassing mystery of the world we live in. "The mystery [*Geheimnis*] is lacking in our Dasein, and thereby the inner terror that every mystery carries with it and that gives Dasein its greatness remains absent. The absence of oppressiveness is what fundamentally oppresses and leaves us most profoundly empty, i.e., the *fundamental emptiness that bores us*. This absence of oppressiveness is only apparently hidden; it is rather attested by the very activities with which we busy ourselves in our contemporary restlessness." This is a kind of temporal diagnosis of the constantly accelerating social processes of our era, as diagnosed by sociology. We distract ourselves from ourselves, we give ourselves no time, we cram our free time full of events. Heidegger's diagnosis sounds almost psychoanalytical, as he claims "the very activities with which we

busy ourselves in our contemporary restlessness" are a reaction to, a denial or repression of, the emptiness inside. This is what Max Weber called the "disenchantment of the world": "Thus the growing process of intellectualization and rationalization does *not* imply a growing understanding of the conditions under which we live. It means something quite different. It is the knowledge or the conviction that if *only we wished* to understand them we *could* do so at any time. It means that in principle, then, we are not ruled by mysterious, unpredictable forces, but that, on the contrary, we can in principle *control everything by means of calculation*. That in turn means the disenchantment of the world." Max Weber, "Science as a Vocation," in *The Vocation Lectures*, ed. and with an introduction by David Owen and Tracy B. Strong, trans. Rodney Livingstone (Indianapolis and Cambridge: Hackett Publishing, 2004), 12–13.

15. See Ludwig Hasler's essay "Die Seele im Alltagsmodus" in the "Die Psyche" issue of *Du* magazine (vol. 7, no. 848, 2014). Here Hasler develops the idea that boredom can be traced back to the increasing isolation and focus on self of the individual who is actually genetically predisposed to be part of a group: "The surfeit of being oneself can also be seen as the soul's reaction to its isolation: when nothing more is going on in the soul than itself, it finds it as boring as the monotonous self-dramatization of the TV studio."

16. Quoted in T. R. Payk, "Störungen des Zeiterlebens bei den endogenen Psychosen," *Schweizer Archiv für Neurologie und Psychiatrie* 121 (1977): 277–285.

17. When tasked with operative estimation of 30-second time intervals, the depressive patients in the initial study (t1) produced on average intervals of 14.55 seconds, that is half as long as specified. This means that for these patients time was expanded to such an extent that already after around 15 seconds they thought that a period of 30 seconds had elapsed. F. T. Melges and C. E. Fougerousse, "Time Sense, Emotions, and Acute Mental Illness," *Journal of Psychiatry Research* 4 (1966): 127–140.

18. T. Fuchs, "Temporality and Psychopathology," *Phenomenology and the Cognitive Sciences* 12 (2013): 75–104.

19. Depressive patients regularly report a "deceleration in the passing of time" or "time standing still." For accounts by patients see P. Hartocollis, *Time and Timelessness* (New York: International Universities Press, 1983). These clinical results correspond with empirical studies showing that depressive patients overestimate temporal durations in the range of between several seconds and minutes relative to healthy subjects: T. Bschor, M. Ising, M. Bauer, U. Lewitzka, M. Skerstupeit, B. Müller-Oerlinghausen, and C. Baethge, "Time Experience and Time Judgment in Major Depression, Mania and Healthy Subjects: A Controlled Study of 93 Subjects," *Acta Psychiatrica Scandinavica* 109 (2004): 222–229. In fact, the significance of the data in time estimation studies in the case of depressive patients is not clear, but this is surely due to the methodological variety of different time tests; not least it is due to the different temporal durations that were tested. For comparison, see a study that used short temporal durations of around one second, which the patients underestimated relative to healthy subjects: S. Gil and S. Droit-Volet, "Time Perception, Depression and Sadness," *Behavioral Processes* 80 (2009): 169–176.

20. T. D. Wilson, D. A. Reinhard, E. C. Westgate, D. T. Gilbert, N. Ellerbeck, C. Hahn, C. L. Brown, and A. Shaked, "Just Think: The Challenges of the Disengaged Mind," *Science* 345 (2014): 75–77.

21. K. A. Mathiak, M. Klasen, M. Zvyagintsev, R. Weber, and K. Mathiak, "Neural Networks Underlying Affective States in a Multimodal Virtual Environment: Contributions to Boredom," *Frontiers in Human Neuroscience* 7 (2013): 820.

22. N. A. Farb, Z. V. Segal, H. Mayberg, J. Bean, D. McKeon, Z. Fatima, and A. K. Anderson, "Attending to the Present: Mindfulness Meditation Reveals Distinct Neural Modes of Self-reference," *Social Cognitive and Affective Neuroscience* 2 (2007): 313–322; N. A. Farb, Z. V. Segal, and A. K. Anderson, "Mindfulness Meditation Training Alters Cortical Representations of Interoceptive Attention," *Social Cognitive and Affective Neuroscience* 8 (2013): 15–26.

23. B. K. Hölzel, U. Ott, T. Gard, H. Hempel, M. Weygandt, K. Morgen, and D. Vaitl, "Investigation of Mindfulness Meditation Practitioners

with Voxel-Based Morphometry," *Social Cognitive and Affective Neuroscience* 3 (2008): 55–61. Of course the insula is embedded in a whole system of other brain areas that show alterations in the case of experienced meditators. For an overview, see B. K. Hölzel, S. W. Lazar, T. Gard, Z. Schuman-Olivier, D. R. Vago, and U. Ott, "How Does Mindfulness Meditation Work? Proposing Mechanisms of Action from a Conceptual and Neural Perspective," *Perspectives on Psychological Science* 6 (2011): 537–559.

24. E. Luders, A. W. Toga, N. Lepore, and C. Gaser, "The Underlying Anatomical Correlates of Long-term Meditation: Larger Hippocampal and Frontal Volumes of Gray Matter," *Neuroimage* 45 (2009): 672–678.

25. N. A. Farb, Z. V. Segal, H. Mayberg, J. Bean, D. McKeon, Z. Fatima, and A. K. Anderson, "Attending to the Present: Mindfulness Meditation Reveals Distinct Neural Modes of Self-reference," *Social Cognitive and Affective Neuroscience* 2 (2007): 313–322.

26. M. E. Raichle, A. M. MacLeod, A. Z. Snyder, W. J. Powers, D. A. Gusnard, and G. L. Shulman, "A Default Mode of Brain Function," *Proceedings of the National Academy of Sciences* 98 (2001): 676–682.

27. G. Northoff, A. Heinzel, M. de Greck, F. Bermpohl, H. Dobrowolny, and J. Panksepp, "Self-referential Processing in Our Brain: A Meta-analysis of Imaging Studies on the Self," *Neuroimage* 31, 440–457; G. Northoff and F. Bermpohl, "Cortical Midline Structures and the Self," *Trends in Cognitive Sciences* 8 (2004): 102–107. See in particular his book: G. Northoff, *Unlocking the Brain*, vol. 2, *Consciousness* (Oxford: Oxford University Press 2013).

28. A. D'Argembeau, D. Feyers, S. Majerus, F. Collette, M. van der Linden, P. Maquet, and E. Salmon, "Self-reflection across Time: Cortical Midline Structures Differentiate between Present and Past Selves," *Social Cognitive and Affective Neuroscience* 3 (2008): 244–252; D. L. Schacter, D. R. Addis, D. Hassabis, V. C. Martin, R. N. Spreng, and K. Szpunar, "The Future of Memory: Remembering, Imagining, and the Brain," *Neuron* 76 (2012): 677–694.

29. A. Berkovich-Ohana, J. Glicksohn, and A. Goldstein, "Mindfulness-Induced Changes in Gamma Band Activity: Implications for the Default Mode Network, Self-reference and Attention," *Clinical Neurophysiology* 123 (2012): 700–710.

30. An initial study demonstrates empirically how, in the case of mind wandering, short temporal durations are actually underestimated: D. B. Terhune, M. Croucher, D. Marcusson-Clavertz, and J. S. Macdonald, "Time Contracts and Temporal Precision Declines When the Mind Wanders," *Journal of Experimental Psychology: Human Perception and Performance* 43 (2017): 1864–1871.

31. S. M. F. Rayport, M. Rayport, and C. A. Schell, "Dostoevsky's Epilepsy: A New Approach to Retrospective Diagnosis," *Epilepsy & Behavior* 22 (2011): 557–570.

32. Fyodor Dostoevsky, *The Idiot*, trans. Richard Pevear and Larissa Volokhonsky (London: Vintage, 2003), 225–226. Here Prince Myshkin's description of the ambivalence provoked by the experience of auras is something between a materialist, neurological explanation and phenomenal ecstasy: "All those flashes, and glimpses of a higher self-sense and self-awareness, and therefore of the 'highest being,' were nothing but an illness, a violation of the normal state, and if so, then this was not the highest being at all but, on the contrary, should be counted as the very lowest. And yet he finally arrived at an extremely paradoxical conclusion: 'So what if it is an illness,' he finally decided. 'Who cares that it's an abnormal strain, if the result itself, if the moment of the sensation, remembered and examined in a healthy state, turns out to be the highest degree of harmony, beauty, gives a hitherto unheard-of and unknown feeling of fullness, measure, reconciliation, and an ecstatic, prayerful merging with the highest synthesis of life?'" (p. 226).

33. F. Picard and A. D. Craig, "Ecstatic Epileptic Seizures: A Potential Window on the Neural Basis for Human Self-awareness," *Epilepsy & Behavior* 16 (2009): 539–546; A. M. Landtblom, H. Lindehammar, H. Karlsson, and A. D. Craig, "Insular Cortex Activation in a Patient with 'Sensed Presence' / Ecstatic Seizures," *Epilepsy & Behavior* 20 (2011):

714–718; F. Picard, "State of Belief, Subjective Certainty and Bliss as a Product of Cortical Dysfunction," *Cortex* 49 (2ǀ03): 2494–2500.

34. V. Monfort, M. Pfeuty, M. Klein, S. Collé, H. Brissart, J. Jonas, and L. Maillard, "Distortion of Time Interval Reproduction in an Epileptic Patient with a Focal Lesion in the Right Anterior Insular / Inferior Frontal Cortices," *Neuropsychologia* 64 (2014): 184–194.

35. In the following discussion of the "self" and "time" in schizophrenia, I am guided by the ideas of the Heidelberg psychiatrist and philosopher Thomas Fuchs: T. Fuchs, "Temporality and Psychopathology," *Phenomenology and the Cognitive Sciences* 12 (2013): 75–104.

36. This passage is quoted in German translation by Thomas Fuchs in his article "Selbst und Schizophrenie," *DZPhil* 60 (2012): 887–901, and is taken from a first-person report of an English-speaking female patient quoted in E. R. Saks, *The Center Cannot Hold: My Journey through Madness* (New York: Hachette, 2008), 12–13.

37. See the recent book by the English philosopher Barry Dainton of Liverpool University, describing the integration of the conscious self in a moment and through time: B. Dainton, *Self: Philosophy in Transit* (London: Penguin Books, 2014).

38. Psychiatrists have documented in books and professional journals the disorders in temporal experience suffered by schizophrenic patients, notably: F. Fischer, "Zeitstruktur und Schizophrenie," *Zeitschrift für die gesamte Neurologie und Psychiatrie* 121 (1929): 544–574; E. Minkowski, "Das Zeit- und das Raumproblem in der Psychopathologie," *Wiener klinische Wochenschrift* 44 (1931), no. 11: 346–384; no. 12: 380–384; W. Meyer-Gross, "On Depersonalization," *British Journal of Medical Psychology* 15 (1935): 103–126; P. Schilder, "Psychopathology of Time," *Journal of Nervous and Mental Disease* 83 (1936): 530–546; T. R. Payk, "Störungen des Zeiterlebens bei den endogenen Psychosen," *Schweizer Archiv für Neurologie, Neurochirurgie und Psychiatrie* 121 (1977): 277–285; B. Kimura, "Zeit und Angst," *Zeitschrift für klinische Psychologie, Psychopathologie und Psychotherapie* 3 (1985): 41–50; T. Fuchs, "The Temporal Structure of Intentionality and Its Disturbance in Schizophrenia," *Psychopathology* 40 (2997): 229–235.

39. F. Fischer, "Zeitstruktur und Schizophrenie," *Zeitschrift für die gesamte Neurologie und Psychiatrie* 121 (1929): 544–574.

40. This account too, in which the continuity of time and the self are reported as mutually disordered, is characteristic: "I am not able to feel myself. The one speaking now is the wrong ego ... When I watch television it is even stranger. Even though I see every scene properly, I do not understand the story as a whole. Each scene jumps over into the next, there is no coherence. Time is also running strangely. It falls apart and no longer progresses. There arise only innumerable separate now, now, now ... quite crazy and without rules or order. It is the same with myself. From moment to moment, various 'selves' arise and disappear entirely at random. There is no connection between my present ego and the one before." This was the account given by a female patient of the Japanese psychiatrist Bin Kimura, quoted in T. Fuchs, "Temporality and Psychopathology," *Phenomenology and the Cognitive Sciences* 12 (2013): 75–104.

41. T. Fuchs, "The Temporal Structure of Intentionality and Its Disturbance in Schizophrenia," *Psychopathology* 40 (2997): 229–235.

42. A. Meyer-Lindenberg, "From Maps to Mechanisms through Neuroimaging of Schizophrenia," *Nature* 468 (2010): 194–202.

43. A. Manoliu, V. Riedl, A. Doll, J. G. Bäuml, M. Mühlau, D. Schwerthöffer, M. Scherr, C. Zimmer, H. Förstl, J. Bäuml, A. M. Wohlschläger, K. Koch, and C. Sorg, "Insular Dysfunction Reflects Altered Between-Network Connectivity and Severity of Negative Symptoms in Schizophrenia during Psychotic Remission," *Frontiers in Human Neuroscience* 7 (2013): 216; A. Manoliu, V. Riedl, A. Zherdin, M. Mühlau, D. Schwerthöffer, M. Scherr, H. Peters, C. Zimmer, H. Förstl, J. Bäuml, A. M. Wohlschläger, and C. Sorg, "Aberrant Dependence of Default Mode / Central Executive Network Interactions on Anterior Insular Salience Network Activity in Schizophrenia," *Schizophrenia Bulletin* 40 (2014): 428–437.

44. L. Palaniyappan and P. F. Liddle, "Does the Salience Network Play a Cardinal Role in Psychosis? An Emerging Hypothesis of Insular Dysfunction," *Journal of Psychiatry & Neuroscience* 37 (2012): 17–27.

45. S. de Haan and T. Fuchs, "The Ghost in the Machine: Disembodiment in Schizophrenia—Two Case Studies," *Psychopathology* 43 (2010): 327–333. See also A. K. Seth, "Interoceptive Inference, Emotion, and the Embodied Self," *Trends in Cognitive Sciences* 17 (2013): 565–573.

46. Caution is always to be exercised when interpreting research findings derived from neuroimaging procedures. All too often in brain research we see the tendency to substantiate the activation of an area with a clearly defined function, as if area X were comprehensively responsible for function Y. If area X "lights up," that is willingly (for example, because it supports a theory) taken as proof of the activation of function Y. In a certain sense, I too conflate many of the findings that point in a preferred direction. We must always keep this confirmation bias in mind when doing research. For me it is ultimately about formulating a hypothesis that nevertheless requires rigorous testing. I am grateful to Felix Hasler for his rigorous position on this matter in our discussions, a position he has also presented thoroughly in his book *Neuromythologie*: F. Hasler, *Neuromythologie: Eine Streitschrift gegen die Deutungsmacht der Hirnforschung*, 3rd ed. (Bielefeld: Transcript, 2013). Felix Hasler sent me an e-mail on this topic on November 2, 2014: "I simply can't take this deluge of fMRI studies on complex cognitive functions seriously any more. Epistemologically they are unproductive and still turned out to be ultimately irrelevant. I fear that in 10 years at the most these kinds of fMRI study will be classified under 'Phrenology' in libraries, where they'll sit gathering dust."

47. It is mostly a case of overestimating specified temporal durations and an operative underproduction of intervals: F. T. Melges and C. E. Fougerousse, "Time Sense, Emotions, and Acute Mental Illness," *Journal of Psychiatry Research* 4 (1966): 127–140; S. L. Rubinstein, *Grundlagen der Allgemeinen Psychologie* (Ost-Berlin: Volk und Wissen Volkseigener Verlag, 1977), 342; O. F. Wahl and D. Sieg, "Time Estimation among Schizophrenics," *Perceptual and Motor Skills* 50 (1980): 535–541; L. Tysk, "Estimation of Time and the Subclassification of Schizophrenic Disorders," *Perceptual and Motor Skills* 57 (1983): 911–918.

48. K. H. Lee, R. S. Bhaker, A. Mysore, R. W. Parks, P. B. Birkett, and P. W. Woodruff, "Time Perception and Its Neuropsychological Correlates

in Patients with Schizophrenia and in Healthy Volunteers," *Psychiatry Research* 166 (2009): 174–183; H. P. Volz, I. Nenadic, C. Gaser, T. Rammsayer, F. Häger, and H. Sauer, "Time Estimation in Schizophrenia: An fMRI Study at Adjusted Levels of Difficulty," *Neuroreport* 12 (2001): 313–316; D. B. Davalos, M. A. Kisley, and R. G. Ross, "Deficits in Auditory and Visual Temporal Perception in Schizophrenia," *Cognitive Neuropsychiatry* 7 (2002): 273–282.

49. W. Perry, R. K. Heaton, E. Potterat, T. Roebuck, A. Minassian, and D. L. Braff, "Working Memory in Schizophrenia: Transient 'Online' Storage versus Executive Functioning," *Schizophrenia Bulletin* 27 (2001): 157–176; J. Lee and S. Park, "Working Memory Impairments in Schizophrenia: A Meta-analysis," *Journal of Abnormal Psychology* 114 (2005): 599–611.

50. We might, for example, read the lecture given by the French physician Eugène Minkowski, who gave a speech in 1930 at the Academic Union for Medical Psychology in Vienna, which was published in 1931 in the *Wiener klinische Wochenschrift*: "When I was studying medicine, our science was still completely under the influence of methodologies from the natural sciences, to the extent that we scarcely dared speak of such as the psyche, especially if we understood by this something other than the basic sensations of color, touch, or hearing. The psychic was dealt with thoughtlessly, from the top down, and anything which did not behave according to the model of an anatomical/physiological, if not physical/chemical formula, was dismissed as 'metaphysics.' There was nothing for it but to withdraw into oneself and think things through on one's own, secretly afraid that one might be making a mistake." E. Minkowski, "Das Zeit- und das Raumproblem in der Psychopathologie," *Wiener klinische Wochenschrift* 44 (1931), no. 11: 346–384; no. 12: 380–384. This attitude is still predominant in some areas of biologically oriented medicine and psychiatry. In around the year 2000, I was supervising a doctoral student in medicine, Nikola Landauer, whose research topic was a psychophysical diagnostic procedure for children with dyslexia. This procedure for measuring the temporal order threshold, that is time consciousness, turned out to be sensitive in the case of children with language disorders. The result was a very good doctoral

thesis. Later, having gained her doctoral degree, Nikola Landauer told me that the professor of medicine with whom she was continuing her further training had dismissed her topic as an "exotic subject." It was simply a psychological topic.

51. One problem in the objective measurement of time perception hitherto, for example the overestimation of temporal duration in the case of schizophrenic patients as described above, is that the results represent an average across a group of patients and suitable control subjects, but the objectively recorded time estimations usually had little or no connection to the clinical symptoms.

52. In a whole series of studies, the researchers demonstrated that the reported effects could be reproduced in both directions, left as well as right, but also vertically (above versus below) and under other conditions, using connected and unconnected squares, and using squares with or without preceding "priming" stimuli: A. Giersch, L. Lalanne, M. van Assche, and M. A. Elliott, "On Disturbed Time Continuity in Schizophrenia: An Elementary Impairment in Visual Perception?" *Frontiers in Psychology* 4 (2013): (281).

53. J. R. Foucher, M. Lacambre, B. T. Pham, A. Giersch, and M. A. Elliott, "Low Time Resolution in Schizophrenia: Lengthened Windows of Simultaneity for Visual, Auditory and Bimodal Stimuli," *Schizophrenia Research* 97 (2007): 118–127; A. Giersch, L. Lalanne, C. Corves, J. Seubert, Z. Shi, J. Foucher, and M. A. Elliott, "Extended Visual Simultaneity Thresholds in Patients with Schizophrenia," *Schizophrenia Bulletin* 35 (2009): 816–825; B. Martin, A. Giersch, C. Huron, and V. van Wassenhove, "Temporal Event Structure and Timing in Schizophrenia: Preserved Binding in a Longer 'Now,'" *Neuropsychologia* 51 (2013): 358–371.

54. L. Lalanne, M. van Assche, and A. Giersch, "When Predictive Mechanisms Go Wrong: Disordered Visual Synchrony Thresholds in Schizophrenia," *Schizophrenia Bulletin* 38 (2010): 506–513; L. Lalanne, M. van Assche, W. Wang, and A. Giersch, "Looking Forward: An Impaired Ability in Patients with Schizophrenia?" *Neuropsychologia* 50 (201): 2736–2744.

55. On the various narratives about psychedelic substances, see the reports in E. Davis, "Return Trip: The Re-enchantment of Psychedelics," *Mind and Matter* 10 (2012): 185–194.

56. R. R. Griffiths and C. S. Grob, "Hallucinogens as Medicine," *Scientific American* 303 (2010): 76–79.

57. W. N. Pahnke, "LSD and Religious Experience: A Paper Presented to a Public Symposium at Wesleyan University, March 1967," in R. C. DeBold and R. C. Leaf (eds.), *LSD, Man and Society* (Middletown, CT: Wesleyan University Press, 1967), 60–85; W. N. Pahnke, "Implications of LSD and Experimental Mysticism," *Journal of Religion and Health* 5 (1966): 175–208.

58. P. Watson, *The Age of Nothing: How We Have Sought to Live since the Death of God* (London: Weidenfeld & Nicolson, 2014), 427.

59. M. A. Geyer and F. X. Vollenweider, "Serotonin Research: Contributions to Understanding Psychoses," *Trends in Pharmacological Sciences* 29 (2008): 445–453.

60. R. R. Griffiths, W. A. Richards, M. W. Johnson, U. D. McCann, and R. Jesse, "Mystical-Type Experiences Occasioned by Psilocybin Mediate the Attribution of Personal Meaning and Spiritual Significance 14 Months Later," *Journal of Psychopharmacology* 22 (2008): 621–632.

61. U. M. Staudinger, "Personality and Aging," in M. Johnson, V. L. Bengtson, P. G. Coleman, and T. Kirkwood (eds.), *Cambridge Handbook of Age and Aging* (Cambridge: Cambridge University Press, 2005), 237–244.

62. C. S. Grob, M. J. Winkelman, and T. B. Roberts, "The Use of Psilocybin in Patients with Advanced Cancer and Existential Anxiety," in M. J. Winkelman and T. B. Roberts (eds.), *Psychedelic Medicine: New Evidence for Hallucinogenic Substances As Treatment* (Westport, CT: Praeger, 2007), 205–216; R. R. Griffiths, M. W. Johnson, M. A. Carducci, A. Umbricht, W. A. Richards, B. D. Richards, M. P. Cosimano, and M. A. Klinedinst, "Psilocybin Produces Substantial and Sustained Decreases in Depression and Anxiety in Patients with Life-Threatening Cancer:

A Randomized Double-Blind Trial," *Journal of Psychopharmacology* 30 (2016): 1181–1197.

63. An entertaining and informed sociological introduction to the history of the sciences' relationship with psychedelics and the narratives in current research can be found in N. Langlitz, *Neuropsychedelia: The Revival of Hallucinogen Research since the Decade of the Brain* (Berkeley: University of California Press, 2013). A German sociologist teaching in the United States, Nicolas Langlitz visited in particular the two centers of psilocybin research in Zurich (Franz Vollenweider's team) and San Diego (Mark Geyer's team), and shadowed the scientists over a period of months. The book also traces the history of the discovery of LSD by the Swiss chemist Albert Hofmann, who worked on synthesizing the substance and studying its effects at the Sandoz Laboratories in Basel.

64. E. Studerus, M. Kometer, F. Hasler, and F. X. Vollenweider, "Acute, Subacute and Long-term Subjective Effects of Psilocybin in Healthy Humans: A Pooled Analysis of Experimental Studies," *Journal of Psychopharmacology* 25 (2011): 1434–1452.

65. W. T. Stace, *Mysticism and Philosophy* (New York: Macmillan, 1960); K. A. MacLean, J. M. S. Leoutsakos, M. W. Johnson, and R. R. Griffiths, "Factor Analysis of the Mystical Experience Questionnaire: A Study of Experiences Occasioned by the Hallucinogen Psilocybin," *Journal for the Scientific Study of Religion* 51 (2012): 721–737.

66. Further development of Adolf Dittrich's questionnaire led to two versions: OAV, with three main factors; and 5D-ABZ with five main factors. See E. Studerus, A. Gamma, and F. X. Vollenweider, "Psychometric Evaluation of the Altered States of Consciousness Rating Scale (OAV)," *PLoS One* 5 (8) (2010): e12412.

67. E. Studerus, M. Kometer, F. Hasler, and F. X. Vollenweider, "Acute, Subacute and Long-term Subjective Effects of Psilocybin in Healthy Humans: A Pooled Analysis of Experimental Studies," *Journal of Psychopharmacology* 25 (2011): 1434–1452.

68. R. L. Carhart-Harris, R. Leech, T. M. Williams, D. Erritzoe, N. Abbasi, T. Bargiotas, P. Hobdon, D. J. Sharp, J. Evans, A. Feilding, R. G. Wise,

and D. J. Nutt, "Implications for Psychedelic-Assisted Psychotherapy: Functional Magnetic Resonance Imaging Study with Psilocybin," *British Journal of Psychiatry* 200 (2012): 238–244.

69. J. C. Kenna and G. Sedman, "The Subjective Experience of Time during Lysergic Acid Diethylamide (LSD-25) Intoxication," *Psychopharmacologia* 5 (1964): 280–288; H. Heimann, "Experience of Time and Space in Model Psychoses," in A. Pletscher and D. Ladewig (eds.), *50 Years of LSD: Current Status and Perspectives of Hallucinogens* (New York: Parthenon, 1994).

70. The contact came about via the neurobiologist Hans Flohr, who was interested in Franz Vollenweider's ketamine research, because Flohr's glutamate-NMDA hypothesis on consciousness can be investigated using these kinds of psychopharmacological studies: H. Flohr, "Sensations and Brain Processes," *Behavioral Brain Research* 71 (1995): 157–161. Hans Flohr approached Ernst Pöppel, who had supervised my doctorate on the topic of "time perception" at the Institute of Medical Psychology in Munich, and Pöppel then alerted me. On my first visit, members of Franz Vollenweider's team included Felix Hasler, who had elaborated the pharmacological characteristics of psilocybin in great detail, the medical doctor Ulrike Grimberg, and the Australian cognitive neuroscientist Olivia Carter.

71. Studies carried out before our investigation showed how fine motor activities were slowed down under the effects of psilocybin: R. Fischer, S. M. England, R. C. Archer, and R. K. Dean, "Psilocybin Reactivity and Time Contraction As Measured by Psychomotor Performance," *Arzneimittelforschung* 16 (1966): 180–185; O. Tosi, M. A. Rockey, and R. Fischer, "Quantitative Measurement of Time Contraction Induced by Psilocybin," *Arzneimittelforschung* 18 (1968): 535–537; a further study showed that perception of short time intervals in the range of milliseconds was not affected by LSD, a finding that we were able to confirm with psilocybin: L. Mitrani, S. Shekerdjiiski, A. Gourevitch, and S. Yanev, "Identification of Short Time Intervals under LSD 25 and Mescaline," *Activitas Nervosa Superior* 19 (1977): 103–104.

72. M. Wittmann, O. Carter, F. Hasler, B. R. Cahn, U. Grimberg, P. Spring, D. Hell, H. Flohr, and F. X. Vollenweider, "Effects of Psilocybin on Time Perception and Temporal Control of Behaviour in Humans," *Journal of Psychopharmacology* 21 (2007): 50–64; J. Wackermann, M. Wittmann, F. Hasler, and F. X. Vollenweider, "Effects of Varied Doses of Psilocybin on Time Interval Reproduction in Human Subjects," *Neuroscience Letters* 435 (2008): 51–55.

73. Summaries of the studies: F. X. Vollenweider and M. Kometer, "The Neurobiology of Psychedelic Drugs: Implications for the Treatment of Mood Disorders," *Nature Reviews Neuroscience* 11 (2010): 642–651; M. A. Geyer and F. X. Vollenweider, "Serotonin Research: Contributions to Understanding Psychoses," *Trends in Pharmacological Sciences* 29 (2008): 445–453. One original study: F. X. Vollenweider, K. L. Leenders, C. Scharfetter, P. Maguire, O. Stadelmann, and J. Angst, "Positron Emission Tomography and Fluorodeoxyglucose Studies of Metabolic Hyperfrontality and Psychopathology in the Psilocybin Model of Psychosis," *Neuropsychopharmacology* 16 (1997): 357–372.

74. F. X. Vollenweider, "Brain Mechanisms of Hallucinogens and Entactogens," *Dialogues in Clinical Neuroscience* 3 (2011): 265–279.

75. H. Flohr, "Sensations and Brain Processes," *Behavioral Brain Research* 71 (1995): 157–161; H. Flohr, U. Glade, and D. Motzko, "The Role of the NMDA Synapse in General Anesthesia," *Toxicology Letters* 100 (1998): 23–29; H. Flohr, "NMDA Receptor-Mediated Computational Processes and Phenomenal Consciousness," in T. Metzinger (ed.), *Neural Correlates of Consciousness: Empirical and Conceptual Questions* (Cambridge MA: MIT Press, 2000), 245–258.

76. R. L. Carhart-Harris, R. Leech, P. J. Hellyer, M. Shanahan, A. Feilding, E. Tagliazucchi, D. R. Chialvo, and D. Nutt, "The Entropic Brain: A Theory of Conscious States Informed by Neuroimaging Research with Psychedelic Drugs," *Frontiers in Human Neuroscience* 8 (2014): 20; R. L. Carhart-Harris, D. Erritzoe, T. Williams, J. M. Stone, L. J. Reed, A. Colasanti, R. J. Tyacke, R. Leech, A. L. Malizia, K. Murphy, P. Hobdon, J. Evans, A. Feilding, R. G. Wise, and D. J. Nutt, "Neural Correlates of the

Psychedelic State As Determined by fMRI Studies with Psilocybin," *Proceedings of the National Academy of Sciences* 109 (2012), 2138–2143.

77. Michael Kometer, one of Franz Vollenweider's colleagues in Zurich, sent me an e-mail on this subject on September 15, 2014: "This discrepancy was a surprise to us too. In fact, a possible explanation might be that psilocybin has a significant influence on blood flow (as serotonergic receptors are located on vessels and psilocybin raises blood pressure), which might in part explain the difference. Apart from that, the form of administration is different. In the case of intravenous administration especially, a very quick, sharp drop in blood pressure can occur. Ultimately, secondary effects of psilocybin might also be responsible for the difference, for example the release of glutamate which perhaps doesn't come into play immediately after intravenous administration as in the case of Carhart-Harris, by contrast with the oral ingestion of psilocybin which is measured after an hour (in our experiments)."

78. G. Petri, P. Expert, F. Turkheimer, R. Carhart-Harris, D. Nutt, P. J. Hellyer, and F. Vaccarino, "Homological Scaffolds of Brain Functional Networks," *Journal of the Royal Society Interface* 11 (2014): 2014.0873.

79. These studies are still considered experimental; their effects have yet to be shown in larger clinical trials with the standard set of three clinical phases. Research on drug effects in depression is even more pressing, since meta-analyses clearly show that standard antidepressant medication (selective serotonin reuptake inhibitors; SSRIs) has only relatively small positive effects without clinical relevance: I. Kirsch, B. J. Deacon, T. B. Huedo-Medina, A. Scoboria, T. J. Moore, and B. T. Johnson, "Initial Severity and Antidepressant Benefits: A Meta-analysis of Data Submitted to the Food and Drug Administration," *PLoS Medicine* 5 (2008): e45. Other pharmacological medication carries the risk of addiction, which is not the case for hallucinogens. For an initial open-label study with psilocybin, see: R. L. Carhart-Harris, M. Bolstridge, J. Rucker, C. M. J. Day, D. Erritzoe, M. Kaelen, M. Bloomfield, et al., "Psilocybin with Psychological Support for Treatment-Resistant Depression: An Open-Label Feasibility Study," *Lancet Psychiatry* 3 (2016): 619–627; an overview of clinical-trial studies using LSD, psilocybin, and ayahuasca can be found

in R. G. dos Santos, F. L. Osório, J. A. S. Crippa, J. Riba, A. W. Zuardi, and J. E. Hallak, "Antidepressive, Anxiolytic, and Antiaddictive Effects of Ayahuasca, Psilocybin and Lysergic Acid Diethylamide (LSD): A Systematic Review of Clinical Trials Published in the Last 25 Years," *Therapeutic Advances in Psychopharmacology* 6 (2016): 193–213.

Epilogue: On Scientific Awakening

1. For a discussion of this topic that has been ignored until only a few years ago, see E. Cardena and M. Winkelman (eds.), *Altering Consciousness: Multidisciplinary Perspectives*, vol. 1, *History, Culture, and the Humanities*; vol. 2, *Biological and Psychological Perspectives* (Santa Barbara, CA: Praeger, 2011).

2. F. Picard and A. D. Craig, "Ecstatic Epileptic Seizures: A Potential Window on the Neural Basis for Human Self-awareness," *Epilepsy & Behavior* 16 (2009): 529–546; A. M. Landtblom, H. Lindehammar, H. Karlsson, and A. D. Craig, "Insular Cortex Activation in a Patient with 'Sensed Presence' / Ecstatic Seizures," *Epilepsy & Behavior* 20 (2011): 714–718.

3. Incidentally, we should mention that this attitude ("Doctors are gentlemen, and gentlemen have clean hands") certainly persists in a lesser way in the world of doctors. In April 2013, in a lecture given at the Institute for Frontier Areas of Psychology and Mental Health in Freiburg, Niko Kohls of Coburg University referred to empirical research on this state of affairs: in 2000, in the highly regarded medical journal *The Lancet*, Didier Pittet and his colleagues described a study into the improvement of compliance with hygiene rules. In this study, around 20,000 instances of hand-washing were investigated. A promotional campaign improved the frequency of hand-washing from 48% to 66%. As a whole, the nursing staff washed their hands most frequently, and the doctors most infrequently. D. Pittet, S. Hugonnet, S. Harbarth, P. Mourouga, V. Sauvan, S. Touveneau, T. V. Perneger, and Members of the Infection Control Programme, "Effectiveness of a Hospital-wide Programme to Improve Compliance with Hand Hygiene," *Lancet* 356 (2000): 1307–1312.

4. O. Blanke, S. Ortigue, T. Landis, and M. Seeck, "Neuropsychology: Stimulating Illusory Own-Body Perceptions," *Nature* 419 (2002): 269–270.

5. Using fMRI, an imaging procedure, the special role of the junction between the temporal and parietal lobes, where the angular gyrus is located, but of other areas as well, could be confirmed. In this, the brain activity of a person was registered by the fMRI as they were able deliberately (but virtually) to detach themselves from their body: A. M. Smith and C. Messier, "Voluntary Out-of-Body Experience: An fMRI Study," *Frontiers in Human Neuroscience* 10, no. 8 (2014): 70.

6. Of course the publication of these research results in a prestigious scientific journal was also due to the fact that the researchers were working with methods of brain research, making it more acceptable.

7. An excellent overview of the empirical findings of the psychological and neural effects of meditation can be found in P. Sedlmeier, J. Eberth, M. Schwarz, D. Zimmermann, F. Haarig, S. Jaeger, and S. Kunze, "The Psychological Effects of Meditation: A Meta-Analysis," *Psychological Bulletin* 138 (2012): 1139–1171; K. C. Fox, M. L. Dixon, S. Nijeboer, M. Girn, J. L. Floman, M. Lifshitz, M. Ellamil, P. Sedlmeier, and K. Christoff, "Functional Neuroanatomy of Meditation: A Review and Meta-Analysis of 78 Functional Neuroimaging Investigations," *Neuroscience & Biobehavioral Reviews* 65 (2016): 208–228.

8. M. Beauregard and V. Paquette, "Neural Correlates of a Mystical Experience in Carmelite Nuns," *Neuroscience Letters* 405 (2006): 186–190.

9. T. Metzinger, *The Ego Tunnel: The Science of the Mind and the Myth of the Self* (New York: Basic Books, 2009).

10. See Nicolas Langlitz's historical and sociological discussion of contemporary researchers working on hallucinogens: N. Langlitz, *Neuropsychedelia* (Berkeley: University of California Press, 2013).

11. The philosopher and best-selling author Richard David Precht graphically summarized his experiences in the realm of academic phi-

losophers thus: "My course work in Cologne got off to an inauspicious start. I had pictured philosophers as fascinating people living lives as exhilarating and uncompromising as their ideas: people like Theodor W. Adorno, Ernst Bloch, or Jean-Paul Sartre. But my vision of bold ideas and a bold life evaporated the instant I caught sight of my new teachers: boring middle-aged gentlemen in pedestrian brown or navy suits." R. D. Precht, *Who Am I? And If So, How Many?*, trans. Shelley Frisch (New York: Spiegel & Grau, 2011), xiv. Is this some kind of "nerd alert"? The following footnote on research into artificial intelligence (AI) makes the same point: "If one were to judge the projects emerging from the early days of AI research, one would have to define intelligence as what interested highly educated male scientists most. At that time it was considered interesting to make computers play chess, perform integral calculus ..., prove mathematical theories, or solve complicated problems in verbal algebra. Tasks that a four- or five-year-old child could perform easily, such as distinguishing between a coffee cup and a chair, walking on two legs, or finding their way from the bedroom to the living room were not considered intelligent achievements. Equally, aesthetic judgments were not counted among the repertoire of intelligent abilities." J. Fingerhut, R. Hufendieck, and M. Wild (eds.), *Philosophie der Verkörperung* (Berlin: suhrkamp taschenbuch wissenschaft, 2013), 54.

12. P. van Lommel, *Consciousness beyond Life: the Science of the Near-Death Experience*, trans. Laura Vroomen (New York: Harper Collins, 2011).

13. A. J. Ayer presented his experiences, among other things, in a chapter titled "That Undiscovered Country," in A. J. Ayer, *The Meaning of Life* (New York: Charles Scribner's Sons, 1989). A concise description of the incident can be found in A. Macfarlane, "A. J. Ayer," *Philosophy Now* 85 (2011): 32–33.

14. First published in *The Spectator* on October 15, 1988, and reprinted in his book, A. J. Ayer, *The Meaning of Life* (New York: Charles Scribner's Sons, 1989).

15. For example, P. van Lommel, R. van Wees, V. Meyers, and I. Elfferich, "Near-Death Experience in Survivors of Cardiac Arrest: A Prospective Study in the Netherlands," *Lancet* 358 (2001): 2039–2045.

16. This information is taken from Rachel Aviv's article "Hobson's Choice," published in the online journal *Believer* (October 2007): https://www.believermag.com/issues/200710/?read=article_aviv. Accessed December 29, 2017.

17. Thomas Nagel, *The View from Nowhere* (Oxford: Oxford University Press, 1986), 10.

Index